Farmer's Boy

FARMER'S BOY
John R. Allan

Illustrated by
Douglas Percy Bliss

Man is a noble animal, splendid in ashes and pompous in the grave, solemnizing nativities and deaths with equal lustre, nor omitting ceremonies of bravery in the infamy of his nature.
SIR THOMAS BROWNE

BIRLINN

First published in 1935
This edition published in 2009 by
Birlinn Limited
West Newington House
10 Newington Road
Edinburgh
EH9 1QS

www.birlinn.co.uk

ISBN: 978 1 84158 822 3

Set in Georgia at Birlinn

Printed and bound in Great Britain by
CPI Cox & Wyman, Reading

TO
WILLIAM MORRISON
In Memory of *The Scots Observer*
and the *Nuits St Georges*

Contents

Illustrations

your aye,

Charlie Allan.

Foreword

By the author's son, Charlie Allan

'You'll never be the man your father was.' I was subjected to that peculiarly Aberdeenshire 'tak doon' often as a boy and as a young man. Perhaps oddly, I was always rather proud of it. My father was a very agreeable, witty and generous person and universally popular. An unlikely hero, he entered the war as a sapper and was demobbed a captain without ever leaving the country. He was persuaded to do his bit for the Labour Party in 1945 and came within two thousand votes of unseating the great Bob Boothby. There was never a more thankful loser. He was a fine player at golf, tennis and bridge. *Farmer's Boy* was published before he was thirty. I always thought that I would have plenty of scope even if I could never be the man he was.

Farmer's Boy is a work of such maturity and perfection in the use of the English language that people thought the author must be old. For more than fifty years after its publication people used to be astonished that my father was still alive.

Which raises the question: why did a talent that flowered so early not leave a bigger literary legacy? Why are there not more than the nine books, including one novel published posthumously?

There are two reasons, apart from the years wasted by the war. The first is John R. Allan's success as a broadcaster, but more particularly as a writer for radio. Between the late thirties and the mid-fifties he wrote hundreds of plays and illustrated talks, many for schools. He even had his own sitcom whose heroes were staff at Dreepieden railway station.

He made what seemed like a lot of money but it took a great deal of John Allan's time. And the trouble with radio is that it leaves nothing for posterity.

In 1952 John Allan published his definitive work, *The North East Lowlands of Scotland*. It is his masterpiece and the masterpiece of that different part of Scotland. Two years later he almost died of pneumonia complicated by cirrhosis of the liver. He never recovered his full vigour and at 49 his career was effectively over. He wrote his potboiler (a popular weekly article for the *Press and Journal*) but he wasn't fit for anything more serious.

His last 31 years were full of frustration as he toiled to finish several novels and the sequel to *Farmer's Boy*, the first part of which, *Brig o' Don Boy*, is published here as a postscript. He just didn't have the strength for more.

To the Reader

This is not an autobiography. It is an imaginative reconstruction of the past, based upon the relations of an old man, an old woman and a young boy. The more distant the past, the more imaginative the treatment: therefore no attempt should be made to identify the characters with men and women so long dead. Nothing would grieve the author more than that, in thus attempting to re-create the spirit of a society which has now disappeared, he should do disservice to any of those whom, though he knows them only by hearsay, he loves this side idolatry—his ancestors.

I

The Family

I was born on the afternoon of the day on which my grandfather signed his third Trust Deed on behoof of his creditors ... the only form of literature in which our family have ever achieved distinction. Though the circumstances of my birth were not such as to cause him any great pleasure, and though his own circumstances were even more involved than usual, the Old Man abated nothing of his usual intransigeance in the face of fortune. When the creditors and their agent had departed, the midwife, an ancient and disillusioned female, presented me to the Old Man. The two generations, I am told, looked at each other in silence across the vastness of seventy years, then the Old Man delivered his grim judgement:

'Gin it gets hair an' teeth it winna look sae like a rabbit.'

And so she left him.

'Aw weel,' said the Old Man, thinking over the events of the day, 'we'd better jist mak a nicht o't.'

He thereupon invited a few old friends, neighbouring farmers long tried in the mischances of this world, to come over that night for a game of nap. They came. They hanselled the new child and the new Trust Deed in the remains of the greybeard and the Old Man collected thirty shillings at nap before morning. Such was the world into which I made my entry on a chill December morning about thirty years ago.

Our family could boast of age if not of honour. We could go back for five generations to one James who was reputed to have been hanged for sheepstealing about the middle of the eighteenth century. All my researches have failed to prove the authenticity of the legend, but I am quite sure that, if James existed, he deserved hanging, whether or not that was his fate.

It was in the first years of the nineteenth century that my grandfather's grandfather came to Dungair. At that time it was no more than a windy pasture between the moss and the moor. By the terms of his lease he had to bring so many acres under the plough within a term of years and make certain extensions to the steading. Nothing at all has survived of this ancestor, beyond the fact that his name was John, that he married a wife by whom he had three sons, and that he died in 1831 in the sixty-second year of his age.

Did I say nothing of his personality has survived? I was wrong, for he took in sixty acres from the moss and the moor, manured and ploughed them and gathered ever-increasing harvests from them. He died in the middle of October and his last picture of this earth may have been the stooks row on row in the Home Field, twenty acres of grain where there had been only rashes and heather before he came. The stooks were his memorial and every fifth year they stood there to his honour, and so will stand as long as harvest comes to Scotland.

John had three sons—Alexander, David and William. David was a good youth, who grew up into a pious man. He was a potter to trade, a great reader in a muddled sort of way and a model to the community in all things. Unfortunately he was left a widow man with one child when he was forty. As there were no unattached female relations available (our family has never bred daughters), David had to engage a housekeeper. Bathia was a rumbustious creature of forty-five who completely altered David's life. She believed in the

Four generations at Dungair

Church, was a connoisseur of funerals, but she could not be doing with books. David's muddleheaded studiousness and respectability appalled her. She set about changing all that. A lot of nasty things were said about her, but I think she had only the best intentions. Anyway she certainly jazzed up David a bit. First of all she gave little tea parties. Then porter and ale parties. Then they had a grand wedding at which only the fact that one of the police was best man saved the whole crew from being run in. David's second go of wedded bliss was not exactly one long sweet song, because there were too many mornings after, but he and his good lady certainly did add to the gaiety of the village in which they lived. And David enjoyed it. When anyone spoke of his first wife he used to sigh—but people were never quite sure that he wasn't thinking of the years he had wasted on her goodness.

William was altogether a simpler case. He was a roisterer from birth. If there was any mischief to do, he did it. If there was blame to be taken, he took it. If there were girls to be kissed, he kissed them. He took no thought for the morrow, but ate and drank and flourished like a sunflower. He was tall, broad and red-faced—a masterful jolly man, fit to be a publican. And a publican he became in a sort of way—at least he also married a widow and together they ran a discreet little alehouse in a discreet back lane. William was one of his own good customers, and thereby achieved his great distinction, which was half a nose.

It happened this way. William had been drinking all morning with his friend Sandy the butcher and, when Sandy announced that he was going back to the shop to cut up a beast, William insisted on going with him. After the usual amount of drunken argument, they decided that William would hold the beast in position while Sandy hacked it up with the cleaver. They got to work. Unfortunately William's legs were not as steady as they might have been, nor Sandy's

4

aim as accurate. No matter whose was the blame. Sandy fetched the beast a mighty whack with the cleaver but missed and sliced off the point of William's nose instead .... Sandy was terribly sorry of course, but William always made light of the accident, though his wife used to say that he never looked the same man again.

Alexander was a better man than his brothers. He had the same physical appearance which persists in the family even to-day—the hard round head covered with thick black hair, the broad shoulders, the deep chest, the rather short legs and the general aspect of an amiable bull. He liked good living, by which I mean dancing, drinking, putting sods on the tops of neighbours' chimneys, and courting every pretty girl within ten miles. He was a roistering young man who seemed destined for the Devil, but he had one great passion which gave him a true bearing among the devious ways of his pleasure. He loved Dungair with a constancy and devotion that he never showed towards any human being. He was extravagant; he was splendidly generous and he was an indifferent business man, with the result that he was always poor, but no matter who should have to make sacrifices it was never the farm. It would have been easy enough for a husbandman of his skill to have cut down the supply of guano for a year or two without doing any great damage or showing the nakedness of the land. He would far sooner have gone naked himself. Good husbandry was his point of honour—a point cherished to idolatry. It was his religion, perhaps the only religion he ever had. It was his one ideal and he never betrayed it. When he came to die he too left his memorial in the sixty acres he took in from the moss and moor, and in the richness which he had added to John's rather grudging fields. He left more than that, for all his descendants inherited something of his care for his beloved acres. John was the pioneer, but Alexander was the Great Ancestor. He died in 1878, aged sixty-nine.

I am loth to pass on from Alexander, for he is the one member of the family who has had anything of greatness in him. It was not only in his devotion to Dungair that he was great. He could go into any company of men as an equal. Remember that he was only a working farmer—not even a bonnet laird—yet he was one of the best known figures in town. He took his Friday dinner in the Red Lion at the same table as the Provost and the Dean of Guild. He carved his portion from the same joint as men who could have bought and sold a thousand of him, because they respected his craftsmanship in the art of life as much as they enjoyed his broad salt wit which never spared them. Of course life was simpler then. Our town was more intimate, more domestic. A man could find his proper place and be at ease in it. It was the golden age of personality and bred a race of worthies—but none was worthier than Dungair. Take him where you like, he was a whole man. His descendants are only weskits stuffed with straw.

Alexander had six children, four sons and two daughters. There had been five others but they died in childhood, none of them surviving beyond eight months. They would not have been unduly lamented, for stock-breeding teaches a man to face life and death with a certain realism. Perhaps their mother wept for them a little when she had time. That's what women were for. Susan is a shadowy figure about whom her children had very little to say. She died when she was fifty, being then a little queer, which is not altogether surprising.

Alexander's sons were John, Francis, Simon and George and his daughters were Jean and Margaret. George, the baby, was a delightful person, a natural of the most engaging kind. If he had been less happy he would have been a poet. As it was, his life was a blameless lyric. He had no inhibitions, no morals, and fortunately no great appetites. His passion was animals. Cattle, horses, sheep, pigs, goats and mongrel dogs were all alike his little brothers. Like a wise elder brother he protected them and like an elder brother he quite frequently thrashed them. But there was no malice in the thrashings and they seem to have understood it. Certainly he never meant to be cruel—a kinder and more benevolent little man never lived—and if he saw anyone maltreating an animal he became almost homicidal. I never met him, but I have heard so much about him from my grandmother who loved him and loved to tell me about him, that I feel as if I had known him all my life. He seldom left Dungair, for he was a little shy of strangers, though perfectly self-possessed if he *had* to meet them. He preferred to stay at home, where he was cattleman, singing as he worked in the byre or fraising with the beasts with whom he was on the most familiar terms. His only incursion into society was his attendance at the dancing class held every winter in the Smiddy barn. He attended that class for twenty-two years and never managed to learn a step, which was very peculiar, for he often used to dance a minuet to his own whistling when things were going by-ordinary well with him. I rather think he went to the dancing class in order to show off his grey bowler hat of which he was very proud, because he ceased attending the class about the time when a family of mice made their nest in it and he became so devoted to them that he could not bear to turn them out. A lot of people, including his brothers and sisters, thought that George was mad. Only his sister-in-law understood that he was beautifully sane. He adored her and was always giving her little presents of flowers. Sometimes, when she had been more than usually kind, his gratitude became embarrassing, as on the day when he presented her with an orphaned hedgehog of an extremely anti-social temper. Of course George had no idea of taking care of himself. One winter night when he was in bed with a bad cold, he became worried about a score of ewes folded up on the moor. He rose at once and, with only a coat over his nightgown, went up to them through the slashing rain. That finished him. He took pneumonia and died in the thirty-ninth year of his age. Everybody mourned for him, especially his elder brother, for he had been a wonderful cattleman and never asked for wages. His sister-in-law mourned for him too, for there were no more flowers, nor any grim unsocial hedgehogs.

Simon was another remarkable character. Like his grandfather he was a pioneer—but in a very different way. He took to learning. Not that he ever became a scholar, but he had ambitions in that direction. He got the same schooling as the others—the privilege of hearing old Mother Kay damn the weather, the crows and himself till he was twelve. Then he was taken home to work on the farm. The good brown earth that the novelists write about took him to herself for twelve hours a day, and after twelve such hours a man had little taste for anything but a chair by the ingle or a walk in the gloaming with a young woman. Yet Simon had the strange impulse to learning in him. Maybe he had the ambition to wag his head in a pulpit; maybe he wished to be a schoolmaster; maybe he just wished to know. Whatever it was, he began to study the Latin when he was twenty, struggling with the rudiments at the bothy fire while his companions played the fiddle or told strong country stories. We must feel very humble, we who get our learning handed to us on a silver plate—when we think of Simon sitting down to unravel the vagaries of the subjunctive, after twelve hours at the hoe. He could never have hoped for any real reward. His must have been a pure love of learning for itself, without any motives of preferment—unless he was like an old shepherd I once knew who collected terms like 'quantum sufficit', 'e pluribus unum', and 'reductio ad absurdum', in order to swear the more effectively at his dogs. Simon did not get very far with the Latin. At twenty-three he married a wife who taught him that life was real and life was earnest

8

when you had to get meal, milk and potatoes for a family of nine. His brother John helped him into a very small farm where he spent a laborious life sweetened only by an occasional taste of the wonders of science as revealed in odd corners of the still very occasional newspaper. He was a kindly, simple man to whom the world was full of unknown but dimly apprehended mysteries, occluded by a voracious and too self-evident family. He died in 1905, aged fifty, leaving behind him eight children who never did much to justify themselves except produce children of the most conventional urban type.

Great-uncle Francis was another family pioneer. He discovered the Trust Deed racket which my grandfather was to carry to perfection. His father settled him in a small farm but did not give him enough capital to make success even remotely possible. Somehow he managed to carry on for fifteen years until his affairs became so embarrassed that he had to sign a Trust Deed on behoof of his creditors. He then found himself cleared of all financial worries, a situation so strange that he died of it. That is all there is to be said about Francis. It required the superior wits of his brother John to see the Trust Deed as a perpetual haven of financial rest. Francis died in 1884, aged thirty-nine. His widow married a plumber in Dundee and was never heard—or thought of—again. He had six children, all of whom went to foreign parts. Only one, as far as I know, achieved any distinction, he having been bumped-off by an almost famous gangster in a speakeasy in Chicago a few years ago.

My grandfather was the eldest son and inherited many of Alexander's enduring qualities, such as his love of Dungair, his contempt for business, his generosity, his philoprogenitiveness and his short strong figure. When I knew him, he was an old man whose sins were coming home to roost on his dauntless shoulders, but they tell me he was a splendid man in his potestatur. When I was a not so little boy we used to drive to church every Sunday, the Old Man and I, in a pony trap drawn by an old white mare. The venerable old gentleman sat high above me on the driving seat with his antique square hat—something between a bowler and a tile—and his square white beard, a ripe old pagan casting a wise eye over the fields he loved so well and the people he despised so truly. The church-going was a rite which he always honoured though he held all religion in contempt, and I am sure those Sunday morning drives through the woods, while the bell sounded so graciously across the shining river, were the pleasantest hours of his life. During the sermon he would lean back in the pew, fold his hands across his stomach, and, fixing the Evangelist in the south-east window with a hard blue eye, would enjoy in retrospect all the wickedness of his diverting life. When the service was over, we went out into the sunlight again, yoked the white mare into the trap and drove home through the woods, while the jingle of the harness mingled so melodiously with the cooing of the pigeons. Every now and then the Old Man would mention some worshipper who had been in church that morning and add a biographical note of which I was too young to understand anything except that it was scandalous. To this day I have never understood why the Old Man went to church, but he may have thought that his presence would be a strong antiseptic against the parish becoming too much infected with religion. On the other hand, church may have been just another place to go to and he was a great goer-to-places.

He had been a famous figure at fairs and markets since ever he became a man. Any fair or any market came alike to him, but his favourite diversion was the Aulton Market, held in the beginning of November on the glebe of St. Machar in Old Aberdeen. The fair was of very great antiquity and even when I was a child it was one of the leading events of the social year. The war killed it as it killed many other ancient institutions. The last time I saw the Aulton Market there were no more than a dozen horses on the field, and no gingerbread and no whisky tents. How different thirty, forty years ago. Maybe five hundred, maybe a thousand horses changed hands that day. The whisky tents seethed with roaring drunken crowds. Great piles of gingerbread and chipped apples (a handful for a penny) melted off the stalls like snow wreaths in thaw; roistering farmers staked their shillings in hopeless attempts to find the lady or spot the pea; fiddlers played reels, pipers piped laments, boxers took on all comers for a guinea, and ballad singers made the afternoon hideous with the songs of Scotland. As the evening came on, gas flares lit up the lanes between the booths, making the shadows yet more drunken as the wind troubled the flames. The town people now came in for their evening's fun—engineers from the shipyards, papermakers up from the Don and hundreds of redoubtable ladies from the Broadford mill. Though the twin spires of St. Machar stood raised like pious hands in horror, and though the tower of King's College maintained her aloof communion with the stars, the saturnalia roared and swirled unheeding on the glebe. And in the middle of it all where the pipers and the singers and the fiddlers were their noisiest you would find Dungair.

The Aulton Market was the scene of his greatest ploy—certainly of the one which gained him the greatest renown. Strangest among the strange creatures attracted to the market was a crazed evangelist known as the Pentecostal Drummer, so named by Dungair because he used to beat a drum at street corners and call on the nations to repentance. Now it so happened that the Pentecostal Drummer had marked down Dungair as a brand specially allotted for him to pluck from the burning. On this particular market day he took to following him round and round the field so that, as soon as Dungair stopped to have a dram or pass a

Aulton Market

joke, the Pentecostal Drummer pitched his stance at his side
and called on him to repent, banging the drum the while.
Not only that, but he had a board on which there was a lurid
drawing of Hell and in red letters—'Beware of the Wrath to
Come.' No matter where Dungair went he found the board
stuck up in his face. There is only a certain amount that a
man will stand even at the Aulton Market. Dungair grew so
annoyed with the Pentecostal Drummer that he suddenly
caught up the board and gave him the weight of the wrath to
come full on the top of his head. The Pentecostal Drummer
showed fight by aiming a kick at Dungair's stomach, but the
Old Man side-stepped, caught the drum in his two hands,
and brought it down with such force on the Drummer's
head that it burst and jammed right over his shoulders. He
then gave the Drummer a smart crack over the ankles with
his staff which so alarmed the poor man, blinded as he was
by the drum, that he let out a hollow booming yell and set
off with a bound, crashing into people right and left, till he
finally came to rest in the ruins of a gingerbread stall. Dungair now felt that he had done enough for honour's sake, and
left the Drummer to square matters with the gingerbread
woman, after impounding his board as a souvenir. Some
time later in the night he made a tour of all his favourite
changehouses, bearing the board aloft, like a banner with
a strange device. And that was how he came to be known
for many a year as 'The Wrath to Come'.

My grandfather was thirty when my great-grandfather
died. As he inherited the lease of Dungair and the headship
of the family, he found himself under the obligation to
take a wife. That can have presented no difficulty, for he
had already a notable reputation as an empiricist. Within
the year he chose the daughter of a neighbouring farmer,
obtained favour with her parents and married her. There
is no record that she loved him, There was even a legend
that she had a romantic passion for a landless youth from

the next parish. And yet I am not so sure, for I once heard her confess, half a century later, that he was a braw man in his big black whisker. It was considered a good match because, if a girl got everything else, she should not expect fidelity. The strange thing is that, however much she had to go without in the hard years to come, she always did get fidelity. It is no business of ours what passion there was between them, nor was I ever privy to their tenderness, for embraces are unseemly on old shoulders and they had an inviolate native dignity. She respected even his faults and he respected the greatness that could respect them. Perhaps there were storms when they were younger, but they were lovely in their age. There was peace in their house when their children left them, and the love with which they cared for me must have sprung from some splendid faith in life, for they were old and could never hope for any reward, nor, as far as I know, did they ever get it.

A few years ago I met an ancient gentleman who used to be our neighbour. He was one of the oldest men I have ever seen, a tiny old man, dried and blanched, like a wand of grass, so that, if the north wind had passed over him, surely he would have been no more. It was a summer day when I went to see him. He was sitting at the window of a blue sunny room, looking over a small garden full of violas and yellow tea roses, and warming in the great sun the tiny silver flame of life that still burned in his ancient body. He knew me as soon as he saw me. I might have been the young Dungair of seventy years ago, he said. Then we fell talking of those distant years when he had ridden a horse and danced at weddings. He mentioned my grandmother. What was she like as a young woman? I asked him. 'The handsomest farmer's wife that ever came to town on market day,' he said, and I'll swear that the life burned stronger far down in his sunken eyes. 'The handsomest wife that ever drove tae market,' he said. We had buried her two months before

14

on a stormy April afternoon. We had expected few people at the funeral, for we had left Dungair, and the family were scattered beyond the seas, therefore it was a surprise to find how many old men had turned out that day. I thought at the time that they were paying their last respects to Dungair and to the family which, the old woman gone, must now be lost for ever in the great world. I was grateful to the old men in their antique coats, and strangely proud that I had been part of Dungair. But now that I saw that tiny gleam in the old man's eye, I began to wonder if it were not the pride and beauty of the young wife that had called the old men to her grave after fifty years.

She was the eldest of a family of eight, born to a farmer and his wife in a small farmhouse nearby. Old Sam, her father, was a kind Christian man who helped to build a church after the Disruption and was an elder in it for many years. Beyond a taste for porter he had no faults except that he could neither make money nor keep his wife in order. Kate was the grimmest good woman God ever made, and I say it who feared her from the first day I saw her. She was rigidly Calvinist, of dominating will and devouring ambition—but I will deal with her again. My grandmother inherited her mother's strength and all her father's sweetness. As the family were poor she had to work hard in the house all morning and at the herding all afternoon. She liked the herding, for then she used to lie under the shade of a tree reading *The Boys of England* and enjoying desperate adventures in strange seas. She became more domestic in her reading in later years, and when I knew her she was a faithful but very critical subscriber to the *People's Friend*, which, incidentally, is a reason why I have a respect for the work of Annie S. Swan which none of my cultured friends have ever been able to discredit.

She was married when she was twenty, after helping to scrub the pots for her marriage feast. In fact she was still running about the close barefoot half an hour before the minister arrived. She found the home-coming to Dungair a mixed business. There was the matter of a husband with bushy black whiskers and queer brothers, but there was also a blessed freedom after the domination by her mother. She had to take her place as mistress of the farm, and how nobly she did it the old men bore witness at her grave.

I had thought to deal with her mother later on, but I can't restrain myself any longer. She was the daughter of an upcountry farmer about whom I know nothing and in whom I am not at all interested. I suppose she was a smart girl, smart enough anyway to attract the notice of the local laird who engaged her as maid to his daughters. Kate was thus promoted from the croft to the Big House, which she accepted at once as her spiritual home. She seems to have been a favourite servant, so much so that when one of the young ladies married a laird near Dungair she took Kate with her as her personal maid. Now all would have been well if Kate had remained in the Big House. She was born to be the housekeeper of a palace and to rule a large staff of servants with a rod of etiquette. Unfortunately her young mistress, having made a happy marriage for herself, tried to do as much for the little maid. Now there happened to be a handsome gardener whose name was Sam. Their mistress did everything she could to bring them together and hinted at the right moment that, if they were to get married, a nice farm was waiting for them. Kate thought it over, decided that a small farm would lead to bigger things, and married Sam. What Sam thought about it is not on record, but he was an easy man, with the fatalism of the true religious. As a farmer he was never the success that Kate wished him to be. He liked porter, he liked discussing vague metaphysical problems, he liked experimenting with plants and seeds, he was one of the first to bring Dutch milk cows to our part of the country—he did everything except concentrate on

16

making money in big enough quantities to satisfy Kate's ambition. Of course Kate despised her neighbours. So did Dungair; but while he despised them as a lot of fools, she despised them for their lack of style. It was her ambition to show them what a real lady was like. Unfortunately she never had enough money. Ambition thwarted expressed itself in domination. She ruled her family, using religion, mother love, and physical force in a combination of devilish ingenuity. Sam stood it all and said nothing. The children paid her lip service till they grew up. Strangely enough it was my grandmother who gave her the greatest shock of her life. When her daughter went to Dungair, I am sure Kate saw herself as mistress of another household and looked forward to bringing Dungair under her capricious rule. She did not find it easy to get a footing there, as her son-in-law was quite her match in any battle of wits or high-class abuse. She got her chance when her dear daughter had a child. Of course Dungair could not deny her right to be present at such a time. She stayed for a month and left only because of a domestic crisis at home. However, she had got her foot in, and, however much Dungair resented her presence, he had to put up with her. A year or two passed. Other children were born. Kate, having none of her own left, thought she would take the new generation under her rule. She tried it, to Dungair's intense resentment, but it was her daughter, hitherto so mild and obedient, who stopped her little game. The scene became a legend in the family. Many versions are told, but the most credible is as follows:

The family were seated at the supper table at Dungair. Kate ordered David, the oldest boy, then aged four, to say grace. He refused.

'It's only sowens,' he said, 'I'll nae say grace for sowens.'

Kate stuck out her square jaw and frowned.

'Ye'll dae as ye're tellt. There'll be nae godlessness as lang's I'm here.'

David glowered back at her.

'I winna say grace; I winna say grace. Ye're nae my mither, onywye.'

Kate lost her temper completely and ground her teeth, as she always did in anger.

'Eh—ye—ye—' she cried, and, taking up the spoon hit the little boy over the knuckles.

Quick as thought, David lifted the bowl of hot sowens, flung it in her face, then began to greet bitterly.

As soon as Kate recovered from the sowens, her temper rose in a flame, and, after the fashion of the time, she would have exacted vengeance on the child, But my grandmother forestalled her. Taking the little boy into the shelter of her arms, she told her mother that, if she ever interfered with the children, she would never be allowed inside the house again.

Kate collapsed into a chair with a fine swoon of mother love.

'That my ain bairn should speak tae me like that.'

Dungair laughed. 'Greet, woman, greet, then blaw your nose an' ye'll be better. There's naething better for pride than a dose o'ts ain medicine.'

Kate's humour changed quickly. She dried her tears and rose up in all her dignity. 'I'm gyaun hame. But mark my words—God'll curse this hoose wi a terrible curse for the ingrateful thing that's been done this day.'

Dungair laughed again.

'Lat Him curse. Lat Him curse. He couldna send's any waur plague than the ane He's takkin awa the day. But a' the same, I doot if even your God'll curse tae order.'

Kate left the house ten minutes later and did not enter it again until my grandmother's sweetness of nature and the knowledge that Kate could no longer interfere made the way for a reconciliation. Kate never forgave Dungair, but treated him with hatred, which he countered with an overt and ribald amusement at her pretensions. And yet I think he always admired the way she kept her God in a bandbox.

She was a most terrifying woman in her old age. She was short and fat. Her face was square, her jaw strong and corrugated with hard wrinkles, like sandpaper. Her nostrils were flat and wide. Pouches hung down under her eyes. Her skin was the colour of old leather. She had an aura of dynamic greyness, as if she had been cut out of living granite. I always sensed that she was hostile, that she would crab my pleasures. On the rare occasions when she came to stay at Dungair she made me utterly miserable. I fell foul of everybody and wherever I went—even in my bed—I felt her cold green eyes boring into me. I felt they were the eyes of an inimical God, that they were my sins that were finding me out.

These, however, are the reactions of a child and, when all is said, I do Kate an injustice. If I had been twenty years older when I first met her I would probably have admired her. She had a great spirit. In her middle age she had a power and dignity that a duchess might have envied. She was generous too. Whenever I visited her she tipped me half-a-crown that I am sure she could ill afford. But I suppose I was afraid of her; and my fear expressed itself in hatred which, of course, I was always rogue enough to hide. Looking back, I know she had magnificent virtues, but as a child I could not appreciate them because of her faults. Yet those faults would have been accounted virtues in their proper day. She was a woman of an older, sterner time when Jehovah was a god of wrath, and parents were little Jehovahs in their homes. I had been reared under a kinder dispensation, and there were too many years between us for us even to be friends.

I will not say anything at this point about the generation that immediately preceded me. They made little impression on myself and less on the farm and therefore do not belong to this story. But the two Johns, Alexander, Simon, George and the rest—these were my ancestors. They were a coarse lot but hearty. They lived with gusto and died in time. I am proud of them, for they went their own way and if the Devil went with them they did not lack for pleasant company.

II

The Little World

My earliest memory of this world is of myself standing in the middle of the big flower bed in the garden at Dungair. It was a quiet summer afternoon. The sun burned like molten gold in the leaves of the big beech tree in the south-east corner of the garden. No doubt a bird was singing. I was a very small, fat, round child clothed in a tight blue jersey and inadequate trousers made down from something cast off by my Uncle Sandy. My big round head was close-cropped, except for a forelock that stood straight up from my forehead so that I looked like a badly snedded turnip. My enormous boots turned up in horns of plenty at the toes; they've always been like that and always will be.

Well, there I was, busy in the flower bed that swept round before the house like a rich carpet spread at an old woman's feet, busy with the best intentions in the world but with the most disastrous results. I was by way of helping my grandmother by weeding in the flower bed and I had just uprooted a dozen violas that were her dearest pride. I had done it very neatly too, disturbed the earth as little as possible and laid out the poor violas in a row to wither in the sun. When I had finished, I threw down my spade, clapped my hands in delight at my good deed, and ran into the house to fetch the Old Woman. She was making ginger-beer at the time—I can still remember the sweet strong smell of the brew—but I insisted that she go out into the

21

garden to see the surprise that I had prepared for her. She went, with the young gardener dancing before her, right up to the flower bed.

'Look, grandma,' I cried, 'isn't it bonny?' pointing to the row of violas already losing their exquisite texture in the fire of the sun.

She looked at the violas—just looked and said nothing.

I looked at her—and waited for praise.

It did not come.

Heaven knows what struggle went on inside her, but she suddenly bent down, picked me up and, still without a word, carried me into the scullery where she gave me such a bathing as I will remember till the day I die.

The incident was never mentioned again.

It bears a remote resemblance to George Washington's celebrated affair with the cherry tree and I have often wondered what would have happened if I had been put to the same test. I would probably have been a little less noble, for I was a child of some imagination and I seemed to know by instinct that, while the naked truth can be told only once and even then with uncomfortable consequences, a well embroidered lie can be told any number of times with adaptations to suit the audience. Lying is an art, of course, and the supreme art of the liar is in knowing when to tell the truth. A good liar is a social asset; a bad liar is a nuisance; and those people who get a reputation for downright veracity are downright nasty, for they tell the truth only when it is unpleasant.

In spite of the usual disappointments, wounded vanity and mornings after, I have always regarded this world as a rather pleasant place. I think the reason is that I spent so much of my little years in the garden at Dungair. It was a wide and sheltered close, with the house against the north, seven tall ash trees and a wall against the east and the cart shed against the west. It lay open to the south and gathered all the warmth of the sun into its arms. Fruit

bushes filled the east border, strawberries the middle and
vegetables the west, while the big flower border swept round
in front of the house. An old apple tree grew in the middle
of the strawberry bed. Sometimes it bore a harvest of sour
unsociable little apples, but more often it was only a choir
for the multitudinous sparrows that swung there gently in
the breeze, like little pirate boats on a quiet tide, waiting
a chance to raid the strawberry beds below. There was an
arbour between the east wall and the house where a small
thatched bower nestled under the golden laburnums. The
bower contained a spacious seat made out of a mahogany
bed and a number of mattresses, where old lady visitors
like Miss Betsy, the sewing woman, used to take their rest
on summer afternoons and add a somnolent burden to the
sweeter humming of the bees.

The garden was my grandmother's own particular world
where she spent every hour she could spare from her duties
in the house. It is thus that I remember her best, an old
woman grown silver grey, a little old woman in a soiled
gown and sacking apron, her nails cracked, her hands torn
and a smudge of earth on her face, moving slowly among
the bushes with a pruning knife, going stiffly down on her
knees to weed the seedlings, or gathering fruit so gently
that her touch seemed a caress. She was not a beautiful
woman poised against an exquisite garden. She herself
was the gardener; she was part of the garden. It was her
greatest joy to care for it and it repaid her most bountifully.
She had had a hard life. She had built many castles in the
quick-sands of human nature and had seen them fall in
ruins. She had tried to pattern life after her own ideas but
it had defied her and delivered her to sorrow. Now in her
old age she had learned wisdom to let human kind gang its
ain gait. She was content to cultivate her garden.

Her love of life was her religion. Just as she had tended
Uncle George's hedgehog so she tended her garden and so

she tended me. The old garden and the young child were all that the years had left unspoiled for her and her love knit us into a trinity that was very lovely in itself and grows the more beautiful in retrospect. Until I went to school the garden, the old woman and the child were inseparable. I played there all day, hunting tigers in the gooseberry bushes, stalking sparrows among the strawberries, or taigling about at the old woman's coat tails; and, when I lay in bed at night, I could hear the wind rustling in the ivy and the everlasting sweet peas tapping at the window pane.

The old woman and I told enough fairy tales in the garden for it to be enchanted, and indeed it was, for every summer brought a mystery. When I was a very little boy and for long after, I always awoke on summer mornings to find a saucer with seven strawberries on my window sill. They were big luscious berries with the dew still on them, picked that hour where they had sheltered beneath the leaves. But who picked them and who put them on my window sill—that was the mystery. I didn't know and nobody that I asked seemed to know either, but there they were without fail. Perhaps the sparrows knew, and that may explain the clamour they raised all morning as they rocked on the barren apple tree flyting at injustice.

The sun did not shine every day then, and sad experience has taught me that Dungair is situated in one of the grimmest parts of Scotland where the cold granite lies at the heart of the soil, but the influence of that garden was such that, when I first went to the West of England, I felt that I had come home.

During the seven months of the year when the sun did not shine I had the old house to play in. It was a wide comfortable place that had been lived in for generations, each of which seemed to have taken its vices with it and left its virtues to hallow its beloved walls. It stood about a hundred yards back from the main road, with the farm

The Boy and his Grandmother in the garden

buildings making the other three sides of a rectangle behind it. This rectangle contained a wide close which was always littered with all manner of agricultural things—clamps of turnips, ploughs, cart wheels, young pigs, barrows, hens and kittens.

The kitchen which opened off the close was a high, rather dark room with a big fireplace, in which you could burn a whole barrow load of peats at a time, and a shapeless funnel of a chimney in which the wind howled most eerily on winter nights. It was an enchanted place at Candlemas when there was a little snow on the ground. The dark came down from the roof where the hams hung, black like the heads of old dead pirates, and in from the peat house in the sinister recess beneath the stair. It was then my grandmother piled the fire with brushwood and the flames crackled valiantly against the menace of the dark, while showers of sparks went dancing up the chimney. The shadows battled with the flames, advancing and retreating in monstrous bounds across the dun red walls. The tin and pewter on the rack glinted like an army terrible in shields and spears and even the figures in the willow-pattern plates strove in remote fantastic war. I used to sit on the brown cheekstone within the ingle, entranced for whole long hours, then, when I looked at the window, there was the roof of the barn, solid, black, immovable, and beyond it the dark cold sky.

On one side of the kitchen there was the scullery, a damp sort of place with a huge whitewashed boiler, behind which lived a legion of immortal beetles. They must have been immortal for, no matter how many we killed, their numbers remained undiminished like the sands of the sea. Walking about on those stupid creatures was the only blood sport old Dungair really enjoyed. He spent many a long evening in the scullery, treading on them with a large inexorable foot. We may have deplored it, but we encouraged it, for he found it wonderfully soothing at the times when the Trust

Deed galled his rather imperious spirit. He even found it a cure for insomnia and the slightly bent old gentleman in his long white nightgown and carpet slippers must have looked like the Immortal Beetle treading his subjects down with a relentless heel.

A small room off the scullery had once been the maids' room. It was very cheerful, as it had a big window looking on the garden. That window was its undoing—or the undoing of the maids, for their sweethearts found it an all too convenient entry. Though Dungair was a man of broad views he fitted iron bars to the window. However, they could not make a cage for love. Thereafter the maids were given a room in the body of the house. Small difference—there was always the roof.

The room off the scullery was now used as a repository for the lumber that gathers in an old house. You might find anything in it except the thing you were looking for. A narrow stair led up to a bedroom and the storeroom. Uncle Sandy lived in splendid squalor in the bedroom. He was a young man at the morose stage, much troubled by intricate love affairs and very touchy about interference by his family, so he withdrew to the dickey into which only his mother was allowed, and only once a week at that, to change his sheets. Of course I went with her and I can assure you it was a great adventure, for the room was always in a most glorious confusion, rather like a bargain basement on a gents' oddments day, after a whirlwind had passed through it.

I had a free run of the storeroom. This was a little box with unpainted walls and a skylight looking to the south. It had a smell—not one of your miserable titillating smells, but a grand odorous compound of all goodness. There was a long broad shelf laden with pots of jam, each covered with a douce white bonnet on which its name was written in my grandmother's wandering hand. There was a big sack of cane sugar, and a small sack of brown, a crate of apples, a box of oranges, bags and boxes of dried fruits, a box of tea, a jar of coffee beans, bunches of dried herbs which used to give a faint green flavour to my grandmother's cakes, a sheaf of that insipid plant called honesty for which I have always had a profound distaste, and a dozen other aromatic creatures. On a summer afternoon when the room grew hot, the many sweets each yielded up their rich incense. The smell was heady, intoxicating, almost tangible. To savour it for half an hour was to feel surfeited forever.

I must not forget the beeswax of which there was great store, for thereby hangs a tale. I have told you that I was born on the afternoon of the day when my grandfather signed his third Trust Deed, but the auspices were not entirely unfavourable, for it was in the same year that the bees came for the third and last time. There were always a few wild bees in the turf dykes around the farm, but one year they seemed more plentiful than usual. Nobody heeded them very much, however. Next year they had multiplied exceedingly, invaded the house, and became rather a nuisance when they mistook the Old Man's face for a sunflower or tried to save time by stealing jam. Still no one bothered to find their nest and with the first chill wind they disappeared. The third summer was one of those rare seasons of ungrudging sunshine when even our lean earth seems to open her heart in flower. The bees swarmed about the house in millions. The garden was full of their low contented hum as they raped the sweetness from the flowers that kept budding in richer and rarer glory as night after night of gentle rain was succeeded by days of brilliant sunshine. The bees became a plague—but still no one discovered their hive. Then my mother, who was near her lying-in, thought she heard a strange noise on the roof as she lay in bed one afternoon. The bees? And so it was. They had found a small hole beneath a broken slate on the roof and so got into the space between the sarking and the lath where they had made their hive. When the summer came to an end and the store of honey was at its greatest, Dungair sent for a bee expert. They stripped a portion of the roof, dispossessed the bees and stole their honey. My grandmother was always strangely moved when she told me how they carried down the honey, jar after jar of sweetness that the wild bees had gathered. I think it appeared to her like a wonder out of the Old Testament, and in her humble way she may have thought that her God had sent the bees to bless her. It had been seldom enough that Heaven had favoured her, but it certainly overworked that summer, for the Trust Deed and myself arrived simultaneously a month or two later.

Let us return to the kitchen. Opposite the outside door there was a long dairy built on to the side of the house and extending about half the length of the back wall. It was four feet below the ground level and was cool and dark, for its tiny windows admitted hardly any light. A broad shelf of stone slabs ran completely round it, bearing the flat basins of milk on which the cream set in a smooth yellow skin. There were usually cheeses and fine lumps of butter lying about and yards of white puddings hanging from a hook in the wall. Sometimes, and especially in harvest, there was a barrel of beer from which Dungair used to draw his evening draught in a milk jug that held about half a gallon. At other times when my grandmother held a brewing there would be a stock of home-made ginger-beer. This was fiery stuff in hot weather, and I remember many afternoons made exciting by an intermittent cannonade in the dairy as the bottles burst with the pressure of the gas. It was splendid ginger-beer when it could be tamed. The only way to deal with it was to have a jug handy. Then you took the bottle to the kitchen door, held the neck of it away from you and screwed out the cork as rapidly as possible. A jet of beer immediately shot across the close, deluging everybody within fifty yards but, if you were really expert, you might save about half in the milk jug. After the froth had gone down you were left with grand ginger-beer that tickled your nose, made your ears crack and gave you the most lovely gollups. But, good as the stuff was as ginger-beer, I think it would have been even better as a fire extinguisher on the pyrene principle. During the war old Dungair said he would never be afraid of the Germans as long as he had a dozen bottles of it in the dairy. It was, he said, mair damnable than dynamite.

On the opposite side of the kitchen from the dairy a door gave into a little hall from which the staircase swung up to the first floor and another door led into the front hall and the garden. There was a third door in the dark corner under the stair which shut off a mysterious passage down into the dairy. The passage was never used as a thoroughfare but as a store for empty bottles and old coats. It gave off a dismal smell of mice and antiquity and was known as the Morgue. It had a story, of course, which told how my grandmother went into it one afternoon when she was alone in the house and found a tramp dead drunk, asleep. He had come into the house unnoticed, had worked his will on the barrel of beer, and then had crept into the Morgue to sleep it off. He was a ferocious, dirty person, but my grandmother did not hesitate. She took him by the heels and dragged him out to the pump in the yard. Having propped him under the spout, she pumped a stream of ice-cold water over the poor man. That called him out of his drunken sleep quick enough, but it scattered his wits even further than ever, if that was possible. He leapt to his feet, let out a roar and streaked off across country through the crofter toun nearby, shouting that the end of the world had come and all Hell had burst its banks. However, that was long ago and nothing half so exciting happened in the Morgue in my time, though a servant, whom I hated, fell down the steps and cracked her head, but that was no great matter for she was half cracked already. I believe she came to a bad end—but there's nothing remarkable in that either.

In a corner of the back hall there stood an institution called the lobby press. It was a general hold-all on whose many shelves there was a quite unique selection of things. It contained everything, even odd sums of money like threepence hapenny, but it was chiefly remarkable for its bottles. One broad shelf held nothing but medicine bottles, filled with the lees of long forgotten prescriptions. When one of the farm boys had a windy colic of more than usual severity, my grandmother would retire into the lobby press where she mixed him a cocktail according to fancy, while he stood on one leg in the kitchen, prepared for the worst. When the physic was ready (sometimes the mixture fizzed and gave off noxious fumes) my grandmother stood over the patient and watched him drink it, with a look on her face that was half scientific interest and half prayer. If the patient was over twenty-one she gave him a dram to put away the taste, if he was under twenty-one she gave him a cup of tea. He then staggered out to the bothy from which he emerged twelve hours later, swept and garnished. It was a kill or cure medicinal, but the boys were grateful; they had strong constitutions and none of them went quite so far as to die of it.

My own experiences of the lobby press were mixed. It contained those horrors of childhood, ipecacuanha wine, castor oil, senna leaves, and Epsom salts. As my grandmother never did things by halves, she used to give me a potent dose of senna infusion mixed with Epsom salts, in case one alone would be insufficient. They were an inexorable combination. However, the lobby press had its compensations, for it was a rich mine of sugared almonds, plums, pandrops, little sweet cakes, dough rings, toy mice, chocolate pigs, and marzipan. As I was already a complete sensualist (which I am still and hope to remain until I die, for I can imagine nothing more horrible than a life of pure intellect and soda-water), the lobby press became the centre of my thoughts, and it had such an individual existence that even to-day I think of it as a definite geographical locality, like Timbuktu on the Rothiemurchus side of Babylon.

The fourth door in this remarkable hall opened into the parlour, certainly the most comfortable room in Scotland on a winter night. The family lived there, but it was particularly old Dungair's. He liked it so much that he had his bed put there. It was a very high bed from which he could see through the tall window, past the golden laburnums, over the garden wall and across the Home Field to the Lea Park and the woods beyond. Once when I had a bellyache through eating too much candy, I lay on that bed, so high that it felt as if I would need a ladder to get down, and watched the Old Man pulling his moletraps in the Home Field, which was showing a fine breer of oats at the time. Whenever he found a trap with a mole in it, he waved his bonnet to me by arrangement, for we had a very ingenious scheme for starting a rabbit farm in our joint interest if we caught enough moles to pay for the first two rabbits. We must have got the skins, or perhaps Dungair borrowed some of his Trustees' money, for we certainly did acquire a pair of black and white rabbits that were said to be pure, though they did not smell like it. We fed them about ten times a day for the first week, once a day for the second week, and then, as they showed no signs of yielding a quick return, we began to neglect them. Somebody took pity on them in the end and set them free, after which the parish enjoyed a plague of piebald rabbits. I was desolated for a day or two by the loss of my capital, but the Old Man said it was maybe just as well; if we *had* made anything out of them thae damned Trustees would likely have taken the lot. That gave me a very unfavourable view of business men which later experience has confirmed.

When Uncle Sandy was away from home or engaged on secret enterprises in the dickey, the old people and I had the parlour to ourselves. Dungair sat in full-bottomed ease in his big chair on one side of the fire, eating apples and spitting the pips accurately at the centre of the flames; my grandmother sat at the other side, half asleep over the *People's Friend,* and I lay on the hearthrug with The Dook, our mongrel dog. It was hot, drowsy and peaceful, and nothing ever happened except bedtime. I am wrong. There was one bad affair. My grandmother had been eating an orange and had taken out her teeth for greater enjoyment. When she finished the orange she threw the skins into the fire with a movement of her apron. Too late she remembered that the teeth had been in her lap.

'O my teeth,' she cried, 'I've burnt them.'

The Old Man never batted an eyelid.

'Aw weel,' he said, quite calmly, 'that's anither gweed fat stot gaen up the lum.'

Let us leave the parlour and open the door into the front lobby. This was a narrow passage from the front door, with the dining-room and the drawing-room on either side of it. The front door opened on the garden, to which you descended by three stone steps. A heavy iron scraper stood on these steps, which Dungair bought at a roup so that he could leave it about for visitors whom he disliked to tumble over as they left on dark nights. Many of them fell head first into the garden, when he would be mildly sympathetic, at the same time insinuating that it must have been the drink.

The grandfather clock stood in a corner of the hall. It had been at Dungair since John's time and I am happy to say that it is now mine, having survived all the financial crises that have diversified my later years. It was (for nothing seems the same now) a stately piece of architecture, like a rococo cathedral, topped by an ornate tower. On its broad and, I thought, kindly face, it bore the painted seasons—Winter with the ploughman returning to his cottage among the snow; Spring with the sower in knee-breeches, after Watteau; Summer with Phyllis dancing to the tune of Strephon's oat; and Autumn with fine ladies in sunbonnets playing in the harvest field. 'Tempus fugit' was drawn in a fine scroll over all, but the flight was a jolly masquerade. I had always an instinct for ignoring the unpleasant side of life, so, when it rained, and there was no joy left out of doors except on the duckpond, I used to sit on the hall table, telling myself stories out of the painted clock, where the sun always shone and the snow never thawed like the end of all beauty. The clock had also a special hand and dial for telling the day of the month, but, as nobody ever remembered to correct it, it was usually about a fortnight out.

The bottom of the clock, where the weights depended, was too good a cache to go empty, so the Old Man kept a special bottle of Jamaica rum there, against the visits of a retired sea captain who lived in a small cottage about three miles away. When Captain Blades came down to Dungair on his tricycle, the old gentlemen retired into the parlour with a bottle of rum and a kettle of boiling water for high discourse on villainy. After they had finished the rum, my grandmother gave them a meal and Captain Blades mounted the tricycle for home. The Captain was a tremendous old gentleman. He was about seventy, but very fresh, stocky of build, with a fiery blue eye and a pointed beard. I think he had spent his earlier days driving clippers through the roaring forties and the instinct had remained with him when his only barque was the tricycle. It was a heartening sight to see him bareheaded, his white hair blowing behind him, his copper nob gleaming and his beard stuck out like a cut-water, coasting furiously down the braes on the tricycle, gybing at the corners with desperate oaths and nautical directions, and leaning back majestically on the straight as if he were one day from Falmouth for orders, with the rest of the clipper fleet behind him, beaten. When he came to Dungair in summer the ritual of the rum was observed with a slight difference. The old gentlemen took their pleasure in the summerhouse and a jug of cold water was substituted for the kettle. There they would sit drinking their rum on warm afternoons while the light, filtered through the laburnums, dyed them a tropic green and gold, and Captain Blades told his outrageous stories of landfalls beneath the Southern Cross. While the bottle died a glorious death, their shouts of laughter shattered the exquisite peace of the garden as if the bawdy gods were carousing in the glades of Arcady.

'Aye, laddie,' my grandmother would say, 'I doot it's time we wis maskin the tea.'

Captain Blades died, as he lived, in drink, and I am sure that, if he won to Heaven, he entered it swearing profusely. He was a terrible man to swear, not in malice but in the glory of words, and in moments of excitement he would ripe the comminatory splendour of seven tongues. I adored him with a holy, an ecstatic reverence. He was very fond of my grandmother, used to give her recipes for scorching eastern dishes and stole cuttings of flowers for her with piratical gusto. If he had only had a wooden leg he would have been perfect, but I am sure he was always too wily to suffer for his sins.

People have sometimes thought to detect an oriental strain in me and one very Saxon young lady, on whom I was trying to make a good impression, told me I looked like a Jew fur merchant. That must be the influence of the drawing-room. I always thought it was something like Heaven on a smaller scale. It was a big room covered with a dull gold-spangled paper and hung with terrible enlargements of family portraits in ornate gold frames. A high mirror, also set in gold, stood on the mantelpiece which was draped with rich brown velvet fringed with pompoms. A long sideboard with a marble top and a high backed mirror filled all of one side and repeated the gold motif in the tracery of its rose-wood front. The chairs too had a yellow streak in them and the carpet was warm, decidedly warm. Only the table, a vast mahogany affair, had escaped the Midas touch. The room must have been very very wonderful when it was new—it was Dungair's wedding present to his wife and I hope she liked it—but forty years had mercifully faded its glory till it had an atmosphere of diffused sunlight even on the gloomi-est winter day as if the ghosts of old dead summers lingered there. It had seen some grand feasts in its time—but I will tell you about them again. I always felt very reverent in it, and I had long interviews with God whom I imagined to dwell somewhere about the great crystal lamp that hung from the ceiling. I once tried to interest the minister in that make-believe, but he took it badly and spoke to my grandmother about it very seriously. However,

'Better a God in the chandelier than nane ava,' she replied, and that was the end of that.

The dining-room across the lobby was the one cheerless room in the house. It had a waxclothed floor, cold greenish walls, and slippy leather furniture that had never been built for comfort. It looked as if it had been designed by the gloomier sort of Presbyterian on a wet Saturday afternoon in March. That alone was enough to make me dislike it, but I had a good reason to hate it—great-grandmother Kate regarded it as her very own whenever she came to stay at Dungair. The more I think of that old woman the more I see her as a terrible piece of work. To begin with, she ate seaweed which she used to bring with her in a big leather bag. As soon as she got settled down and had her tea, she heated the poker red hot then thrust it into a handful of the seaweed which gave off clouds of smoke that smelled like burning rubber. When the stuff ceased hissing, she gobbled it up, chewing handfuls of it ruthlessly in her iron jaws and crunching, with great satisfaction, the small barnacles that adhered to it. The performance horrified and fascinated me, and, as I heard the barnacles go scrunch between her jaws, I shuddered, for she was so like something malevolent out of a fairy tale. I think our dislike was mutual. You see, my close companionship with my grandmother had made me a friendly and inquiring child. Unquestioning obedience was never in my nature (I hope it never will be). I always wished to discuss an order before carrying it out. So when Kate issued a command I prepared to talk it over as man to man. The very idea, the godless presumption, the unparalleled impertinence infuriated her. She used to swell up with rage, clench her fists and splutter at me. She would have dearly loved to give me a thrashing on the spot, but she did not dare, so she had to content herself by threatening me with all the mysterious terrors of Heaven. Now I had got a very clear idea of Heaven from my grandmother and knew it was something like the drawing-room, set in a limitless garden where a kind old gentleman reigned with the sun for a crown and the beauty of the moon and the stars on his shoulders. So I was puzzled when Kate threatened me with her sadistic old Jehovah. It was quite obvious that our Gods were different. I thought it over for a long time and, as I knew that there was only one God, I came to the only possible conclusion. Well, one afternoon I was playing my favourite game of Trains in the garden. This was a hearty game of rushing along the walks at full speed, kicking up pebbles and blowing a trumpet at the corners. Now Kate was dozing in the summerhouse and the trumpet annoyed her. She called me over.

'Ye'll stop rushin aboot an playin that tooteroo at once,' she said. 'How dare ye mak a noise like that on the Sabbath day? It'll be a wonder if God doesna strike ye dumb.'

I gave a derisive toot.

'Your God couldna.'

She grew terrible like a storm cloud. 'What?'

I tooted again, defiantly. 'Your God couldna. Mine winna lat 'im. Your God's only the Deevil.'

The storm broke—but I had disappeared into the gooseberry bushes. Kate went half mad and, as she did not dare touch me herself, tried to invoke summary justice from my grandparents. She met with a very cold reception. My grandmother told her quite plainly that it was wicked to paint God in such colours, while Dungair said that, as far as Gods were concerned, he saw little to pick and choose among them, but that the bairn was probably right. Thereafter Kate swore that I would come to a bad end—but she never threatened me with the vengeance of Heaven again.

Dungair held his card parties in the dining-room. He was extremely fond of card games, especially of nap, and at least once a month all through the winter he would have a few of the neighbours in for a game. These parties usually began about eight. The five or six men took their places round the table, undid their waistcoats, arranged their small change before them and settled down for the night. My grandmother drew the heavy curtains, heaped up the fire with coal, placed a tray of sandwiches on the sideboard and the greybeard in a cool corner and left them to it. Often and often they played penny nap until breakfast time, at a grievous expense of whisky which Dungair recouped with his winnings, for he had an acute card sense even in a state of drink taken. When I was old enough, I was allowed to watch the games, and before I was ten I developed a mania for cards which lasted until a three years' run of the most appalling bad luck at bridge cured me for ever.

Dungair was not a house with a lot of books, for our family have preferred to live romances rather than read them, but such books as we did own were kept in the dining-room press. There were a few romances by Sir Walter Scott, which I could never abide until the parochialism of modern Scottish fiction and Stevenson's pernickety glamour revealed to me the majesty of *Waverley's* larger air. There were a few stories by Dickens, a Bulwer Lytton or two, some bound copies of magazines and such other odds and ends. The rest were highly coloured love stories of the James Payn school, with names like *What He Cost Her*. I should have thought that the women of Dungair would have had little to learn from books like those.

Now let us go upstairs. The bedrooms, of which there were four, two in front and two at the back, had not the character of the living-rooms. They were just places to sleep in—except the spare room. I have always marvelled at people who could sleep there—I would never have dared. It was a sunny room with a dormer window overlooking the garden, but it had such an air of state that I always lowered my voice when I was in it. The bed was a world's wonder, all gleaming brass and frilly counterpanes, all cold linen sheets and airy eiderdown, ceremonial and severe. When I thought of the friendly chaff bed into which I used to burrow like a little mouse the spare bed used to make me shiver, and it always reminded me somehow of the inconsolate wastes about the Pole. The chest of drawers, dressing-table and wardrobe were of a delicate rosewood and always smelled of apples, so that I called the room the Garden of Eden ... but I doubt if our First Parents would have dared fall in that bed.

The other front room had been the nursery, but my grandmother slept in it herself after the children left home. I too slept there in a small cot until I was six, when I moved in with my grandfather to keep him company in the parlour downstairs. I remember very little about that room. I could sleep well and long in those days and, if I had dreams, they were forgotten by morning. All that I do remember is the miracle of the strawberries and the way the birds sang in the ash trees on summer mornings.

The maid—by which I mean the long succession of magdalenes and half-wits that did the heavy work about the house—lived in one of the back rooms. Of course it was not considered necessary to give a kitchen wench a decent room—she wasn't accustomed to it and wouldn't have known what to do with it. A creaky bed, a cracked mirror, and a rickety table were all she deserved and all she usually got ... a hole into which she could creep at night and from which she could emerge at half-past four, eager for another day's work. Now my grandmother was not of that school of thought, but she was not a revolutionary either and, though the maids' room had some amenities such as a wardrobe and a chest of drawers, it was by no means a Paradise in which a lonely girl might be soothed to sweet slumbers. It was long and narrow with a skylight opening on the north. The walls were distempered a cold blue. There the domestics spent their dreary nights diversified with spasms of bucolic love at the week-ends. If I were a woman, only the direst necessity would make me take service at a farm, for I will be very much surprised if the conditions are one bit better to-day than they were twenty years ago.

The other back room became mine when I was ten, and it was there that I tasted the fearful joys of independence.

That was not all the house, for beyond the scullery there was a tool-shed which you entered from the close. It was dark, dusty and littered with the accumulation of a hundred years. I can't even begin to enumerate the things that had collected in it, or the smells that rose from its earthen floor, for everything had been kept in it, from guano to dogs. It was full of bent, battered and broken tools, but Dungair would never allow any of them to be thrown away. They would make good trade at his roup, he said, and I should imagine that they fetched ten shillings in 1919, that wonder year. There was a drying loft above the tool-shed, full of empty boxes and black shadows. I can't remember that anything ever happened there, though with a family like ours you never can tell.

And that was the old house of Dungair, wide, comfortable, well settled down into its native earth, with its two sets of six chimneys, its dormer windows overlooking the garden, its gilded weathercock on the west gable end and the lovely garden lying before it in the sun. It was more than a house: it was a century of hard lived experiences, to which I was the latest heir. Perhaps I have written a great deal about those experiences and very little about the small boy who inherited them, but those experiences made the boy, and the house was as much his mother as if it had borne him in its womb. It is to my shame that I once despised Dungair. The University and its society of cheap wits seemed to offer a noble culture and an ampler humanity. I climbed the steep ascent towards a morning coat and an English accent, and managed to wear both with distinction. I cultivated philosophies, took part in movements, earned a good income, saved money and was a credit to society. But it was no good. I was merely one among a million climbers, parasites that had their roots in thin air and aspired to a place in the sun. We were lost in a world of words and pretensions. We had no reality. So I said goodbye to the morning coat—I sold it for exactly half what it cost me—and returned in all humility to the place where I began. I am poorer than my grandfather was, for I have only the grandfather clock to deed in trust, but I have at least one thing that is worth all the gentility and aspirations I forswore—I have one tremendous reality behind me, the tradition and loyalties of my fathers.

III
Unwillingly to School

Our family motto has always been 'You take care of yourself and your creditors will take care of your debts.' It is a pleasant philosophy, but a lot depends on choosing the right sort of creditors. In nothing were my family more fortunate than in the people they owed money to. We were always hopeless in financial matters. John the First, Alexander, Francis and my grandfather were excellent farmers but wretched business men. No matter how good the times might be, they spent their harvests before they reaped them. Old Dungair was the worst of all, and he would have lost his farm if it had not been for that blessed institution—the Trust Deed.

The Trust Deed cannot be too highly praised, for under its beneficent provisions his creditors not only took care of his debts but they took care of himself as well. The ritual was as follows. He would enjoy his riotous courses for ten years or so, getting a little more in debt each year, until he was worth about five shillings in the £. When things got so involved that even he began to find them uncomfortable, he would go to his chief creditors and suggest that he assign them all his property under a Trust Deed. The creditors would agree. They took over the farm, appointed him as manager at a reasonable wage, set Uncle Scott, a hardy old relation, over him as trustee and hoped to get back their money somehow. The Old Man and Uncle Scott worked together very well. Dungair managed the farm with more than his ordinary skill, because he had no money for going places, and Uncle Scott saw that the crops were sold to the best possible advantage. It was a good farm. There was money to be made in it. After seven years the creditors would have received maybe sixteen shillings in the £. They would then agree to make the final settlement, the Trust would be wound up, and the farm handed back to the Old Man as good as ever. Then he would have seven more glorious years, followed by another Trust Deed—and so the game went on.

It was not always riotous living that brought Dungair into difficulties. Once it was the foot and mouth disease. Scientists have not yet discovered the cause and treatment of this plague, but their rigorous policy of segregation and slaughter has confined it to sporadic outbreaks, which can be controlled with comparative ease. Forty years ago, however, foot and mouth disease was a nightmare that haunted every cattle farmer. He never knew when some new beast would bring home the plague from market, infect his herd and probably ruin him. Every few years the disease swept over Scotland—and once it came to Dungair. There were suspected cases at some of the neighbouring farms and Dungair was alarmed, for he had bought four new cows at Friday's market. He examined them half a dozen times a day but they seemed all right. Then one morning he noticed that two of them were slavering badly. He sent for the vet. The vet said foot and mouth disease beyond all doubt. So that was that.

There were forty dairy cows in that byre, perhaps a dozen stirks in the small byre beyond, and five horses in the stable. Well, the sanitary official came, with wood and tar and paraffin. The farm boys built great pyres at his orders down in a hollow of the Home Field. When the pyres were ready, they led out the cows, one by one, down to the hollow, shot them, and hauled them on to the pyres; then the stirks and a little calf born that morning; and the quiet lovely Clydesdale mares last of all. It was evening before they finished, and it was growing dark when Dungair laid the first match to the wood. It was a dry windless night in the beginning of November and the world lay dead under the indifferent stars as the fires rose in the hollow and the smoke swelled over the land, heavy with the fat of slain beasts. The hollow boiled and bubbled like the depths of Hell, hour after hour. The smoke infested the house till those within it grew sick. The loss of so many lovely creatures, and the utter ruin of so many dear ambitions, hung over the farm in darkest tragedy and my grandmother wished that she might die that night. And the unbearable rich sweetness of the burnt flesh haunted the house like plague. Morning brought a west wind. The smoke dissolved away, the fire burned out, and the wind scattered the ashes in the hollow.

In the peace of her last years, when she was secure in her small house in a garden by a river, my grandmother would sometimes speak of those terrible days when the smell of burning flesh came down the wind, and the fires would glint on the hilltops all night in witness that tragedy had overtaken yet another farm. A thousand pounds of capital would be lost in a night. After a lifetime of desperately hard work, people would find themselves back at the very beginning. So it happened at Dungair. It was a total loss. There was no compensation, unless in the comforts of religion. 'It's sometimes gie hard tae believe in the goodness o' God,' my grandmother once said, after she had been telling me of those days, 'an easy seat maks an easy faith, an' I some doot the ministers are ower sure o' their seat in Heaven.'

Dungair had not that kind of faith. I doubt if he has got to Heaven, or that he is comfortable if he is there. But he came through all his troubles with a hearty curse for the tricks of fortune, kept his hair and his teeth to the last, and died with a laugh, regretting nothing.

Well, when I was five and ready to go to school, he had enjoyed five years of his third Trust Deed. Things were going well. The bees seemed to have brought a change of fortune. It was not that prices were particularly good in 1911, nor that the world was full of hope, for the seeds of the present chaos were already springing in fields where they had been so carelessly sown, but Dungair had the luck on his side at last. He raised good harvests and brought them home in splendid order. The cows milked well and took no ill. He bought and sold at the right time, turning that extra shilling which makes the difference between profit and loss. When Uncle Scott came to balance the accounts each year, there was always a dividend for the creditors. Three or four years and the Old Man would be his own master again. He had just celebrated his seventy-fifth birthday. His enjoyment of life was unimpaired. He looked forward to seeing a lot of ferlies before he died. And, by heaven, he saw more than he ever bargained for.

The day I went to school brought order and responsibility into a life that had been free as the wind's will in the garden at Dungair. I looked forward to it with a complete lack of enthusiasm. I was an only child who knew little of the ways of other children, and disliked intensely what he did know of them. Such small boys as I had met seemed to think it the height of fun to climb up trees for the pleasure of falling down, whereas I, being extremely fat, had great difficulty in climbing at all. Then they could not understand that the colours of a viola were so rich that they seemed to live, and would shrink from your hand if you touched them, nor did they care at all for the strange pattern of five crows on a bare bough. They trampled the violas and threw stones at the crows. I was woefully backward in all manly sports. I couldn't throw stones for toffee or anything else, but, though the disability has brought me a lot of shame in my time, I have never really missed the accomplishment.

I was very clumsy with a ball, being too slow to run after it and almost unable to catch it in the air. In fact I wasn't a proper boy at all, only an old woman's pet, but everybody agreed that the school would make a man of me.

The process began about eight o'clock on a cold spring morning. I washed and dressed myself like a lamb before his shearers, dumb with apprehension. Then Sally put on my new collar, one of those celluloid things you wash with a sponge, packed my bag with sandwiches and led me away on the mile walk to the school. I marched valiantly at her side along the whin road through the fields to the wood, but, when we got about ten yards into the wood, my spirits failed completely. I was leaving all I knew and might never see them again. I wept as one who had made a three years study of weeping. I lay down among the withered leaves and refused to be comforted. However, Sally knew me. She just let me greet, and, when I had tired myself out, began to tell me a story about a giant. She dangled that giant before my goggle eyes so adroitly that we arrived at the school before I knew what had happened to me. Then she handed me over to Miss Grey, gave me a kiss behind the door, which heartened me wonderfully, and left me alone in my strange new world.

I do not remember very much about the next seven years in school, certainly nothing of any great interest. I was an intelligent child, always a mile in front of the other children. As I was good tempered and lazy I never got into mischief. I was a perfect pupil and no matter what storms raged in the school—they were frequent and terrible—they never broke on my shoulders. It was not until I went to a town school and met children far cleverer at their books than I was, that I had to use my wits to escape the consequences of my laziness. Things began to happen then as they have been happening ever since, but the first seven years passed uneventfully. I learned to read and write with fluency, to do mental arithmetic with incredible speed, so that I could still be quite useful in a draper's shop; to recite portions of the Scriptures with accuracy if not with devotion, to trace maps of Australia with the deserts in the right places, to draw empty bottles that were almost recognizable, to deal promptly with the recurring decimal, and to have clean hands and a shining face. I also learned to play most games with the inconspicuous incompetence which I still retain and to despise all women, which, unfortunately, I soon got over.

I think I was fortunate in my first school. To begin with, it was most picturesquely situated at the edge of a wood on a hill which overlooked the town and the sea. When you came out of the tangle of the wood, you saw the town lying in the valley below you, very beautiful with its towers and steeples shining in the sunlight. The school might have been a bandit lair from which you could sweep down on misty nights to spread murder and rapine among the citizens. We must have been not unlike a gang of young bandits. There were about fifty children in the school, drawn from the farms and cottar houses of the district, mostly the gets of very poor parents and all in a state of disrepair. Our clothes were always torn, our faces dirty and our manners barbarian. Some of us came from what might have been called good Christian homes, some of us even went to Sunday school, but most of us were familiar with the name of Christ only as an oath of common usage. Especially the cottar children who existed in a state of nature. They were the nearest I have known to Rousseau's noble savages, except that there was nothing at all noble about them. I suppose they were a social problem. Their parents were nomadic, unable to settle long in any one place, but changing their employment at least once a year. They spent their lives shifting about from one cottar house to another, and, as they had no sense of continuity, they did not cultivate the domestic arts; their houses were hovels, camping grounds; their gardens lay waste and their whole lives were makeshift. They multiplied abundantly, having only the most primitive notions of birth control. However, those devices were not used until the two rooms of the cottar house were crowded with a father, a mother and seven or eight children, so that there were always plenty of little barbarians to fill the desks in the school.

There would have been about twenty or thirty children between the ages of five and nine, sitting two and two in the pitch pine desks in the Infant Room, where we struggled all day with the mysteries of simple addition and the black magic of longhand. I remember the smells best of all, because they were the most definite things in the room. There were the smell of the desks, of the chalk, of the tin flasks of cocoa heating round the fire, of the rags with which we cleaned our slates, of old wet clothes drying in the close heat, all blended into something musty, like the smell of tunnels with a faint undertone of ammonia. Not all the winds that blew in at the windows from the tulip fields of Holland across the salt blue sea could ever sweeten that scholastic air. You can imagine what a blessing it was to get out at the meal hour, to play in the firwood and to hide among the dried leaves under a broom brush, and how unwillingly we returned when the too short hour was over.

At least we could get clear of it when we were fourteen. What about Miss Thom, the infant mistress, who had spent thirty years there, and had the prospect of spending another twenty? I suppose she had resigned herself to her fate. Perhaps she liked the work. I don't know. But she was a good teacher, had infinite patience, never frightened us, and managed to teach something to even the most stupid, even if it was only to keep their noses clean in class—and that, I can assure you, was comparative civilization for many.

The Infant Room was a cloistered haven from which we could hear Miss Grey raging at the godless multitude in the Senior Room. Many a time when we were due to go into the Senior Room at the term, we used to listen to Miss Grey storming on the other side of the glass partition, while our knees trembled at the prospect of the wrath to come. We used to get a taste of it when she made a raid into the Infant Room. The slate pencils usually began the trouble. The school would be still on a warm afternoon when all the classes had been set their tasks. Then the screech of a slate pencil would tear the silence—screech, screech, screech. If Miss Grey happened to be feeling a little raw the sound would infuriate her. Through she would come with her grey eyes like granite, fix the sinner with a fierce stare and denounce him as a limb of Satan. Such interference was more than Miss Thom could stand. She would request Miss Grey to address all complaints to her as she was in charge of the room. Now two women cannot work day in day out in the nervous atmosphere of a small school without a certain amount of mutual annoyance, and, at an opportunity given, the two ladies would relieve their feelings. Miss Grey became fiery. Miss Thom remained freezingly polite as her temper rose. Their voices clashed like swords, while the whole school looked on with wildly beating hearts. Then, just as things were hotting up nicely, one would suggest that they go outside to talk it over. They went, their voices dying along the passage as they retired to the saner air out-side. After a few minutes, in which the school surmised wildly as to the progress of events and the more daring boys scouted warily along the passage, the ladies would return to their rooms and the school would gradually go to sleep for the rest of the afternoon. Such incidents made a wonderful diversion, but they did put the fear of death into the hearts of those who were due for elevation to Miss Grey's sphere of influence. I contemplated that translation with the utmost foreboding because I felt that Miss Grey would not be prepared to talk things over as man to man in a spirit of sweet reasonableness. Nor was she, but we became great friends in spite of it, and I have the honour still to be her friend.

The fiery humours of Miss Grey enlarged my ideas of the great world, but they did not make as great an impression on me as the dreadful episode of Uncle Sandy and the young pig. The whole business was most deplorable. It gave me a shock that disrupted the foundations of my little world. And it hurt me in a way that I resented most bitterly.

It all arose out of the inquiring temper of my mind. You see our breeding sow—which really belonged to Uncle Sandy—had just farrowed a litter of pigs. That event in itself did not interest me greatly, for Old Dungair was not the man to tell me any tales about gooseberry bushes, and I was enlightened on the whole process of reproduction by one who had some claims to be called an expert. But the little pigs did fascinate me. I leaned over the sty for a long time, and the more I looked the more I wanted to play with them. Impulse and action went hand in hand in those days, so I climbed into the sty and began to play with the friendly little beasts. So far so good. Then my inquiring mind played the Devil with me. I wondered what would happen if I threw a little pig into the air. Would it or would it not fall on its feet like a cat? Then I thought that, as the pigs were very young, they might not be very good at turning, therefore it would be better to throw them as high as possible so that they would have all the more room to turn in. When I had worked it out I decided to put my conclusions to the test. I lifted a little pig and threw it as far into the air as I could. That wasn't very far—not nearly far enough—for the little beast fell back into the straw with a horrid squeal which moved my amiable heart. Unfortunately it touched another heart which was not quite so gentle, for Uncle Sandy came into the shed at the wrong moment and saw one of his valuable pigs whirling towards the roof. He let out a bellow of rage and fell on me

Uncle Sandy

hip and thigh. Hip and thigh is the expression, for he did not listen to my scientific protestations, but hauled down my slacks with a ruthless hand, so that the buttons flew in all directions, laid me across his knee and let me have it. All that could be heard in the shed for the next minute was the impact of a horny hand on tender buttocks, and roars of anger from a young scientist who had been hoist with his own petard. When he had avenged the little pig, Uncle Sandy gave me a final grand smack that echoes in my head till this day, set me down on my two feet and left me there. And there I stood, my world about my ears and my little slacks about my feet, bawling out my anger and humiliation to the unheeding afternoon, while the little pigs rooted and squealed in the pen nearby. I don't know whether my mind or my buttocks ached more. I was a wailing excoriated mass, incapable of thought. After a little I began to move towards the house, greatly impeded by my sobs and the trousers hobbled round my ankles.

Every now and again I stopped for a grander roar, then moved on when I had exhausted it. I had got as far as the middle of the close when my anger, humiliation and pain finally overcame me. I could go no further, but stood there in the cold grey afternoon, with my knuckles crammed into my eyes, my trousers draggling behind me, and my pink shirt fluttering round my tender buttocks. I may have stood there howling for hours, unable to move, until Sally came in from the fields and found me. She asked no questions nor did she offer me any sympathy. She took in the whole situation at a glance, pulled up my trousers, tucked in my shirt, produced two safety pins and so repaired my shame.

'Noo than,' she said very softly, 'what are ye greetin aboot? Come on an' help me tae bake bannocks.'

While we were baking the bannocks, we talked over the business of the little pig, and she managed to persuade me that I had been rather misguided, and that there was a lot to be said for Uncle Sandy. The Old Man settled the matter by treating it as a good joke, and Uncle Sandy and I resumed our normal relations of cheerful enmity. But, and I will say it again, the inquiring mind got a decided set-back from which it never fully recovered.

It had another misadventure about this time, with less painful results, however. My grandmother always raised a lot of poultry, which must have been a great help towards the housekeeping, though the Old Man, like every other farmer then, despised poultry as parasites and worse. I used to help my grandmother with the hens, particularly at hatching time, for I was fascinated by the miracle of the chicken in the egg. The old woman had a trick of placing the egg to her ear, and of being able to tell whether or not it would produce a live chicken. Those eggs that were infertile she placed aside, and I was allowed to break them against the midden wall, a much cherished privilege. Well, one afternoon when I was making a round of the nesting shed alone, it occurred to me that I too might test a few of the eggs. So I unseated a friendly old hen (as a rule the brooders were very kittle cattle) and sounded the eggs. I did not seem to have the knack of it, for though I held them to either ear, I could not decide one way or the other. The passion of scientific inquiry now ran away with me. I just had to know what was inside the eggs. I employed the traditional method. It was great fun—at least to me—but it may have been otherwise to my grandmother who came in to find me breaking up the last nestful, while the dispossessed hens stood disconsolately in a gigantic omelet on the floor, lamenting dolefully. I am told that I looked up with my most innocent round-faced smile and said:

'I wis jist seein' if there wis ony chuckies in them.'

My grandmother looked sadly at the ruins and thought of all the chickens she had counted before they were hatched.

'Aw weel,' she said, 'we fairly ken noo.'

That was all she ever said about it, but I did not need the business discussed as man to man. The look on her face told me that the spirit of scientific inquiry must be tempered with discretion. Life becomes very difficult about the age of six.

I now began to go about the greater world, especially with the Old Man. He liked to have me with him and to tell me the history of the countryside—not the history of cromlechs and cairns, for he wasn't at all interested in that sort of thing, but the personal and far more amusing histories of the people we met, of the women they married and the women they didn't, of the crops they raised and the fortunes they spent, of olden markets and change houses, of long-forgotten battles and ancient lunacies. It was more than gossip and more than the haiverings of an old man, for Dungair could keep a shut mouth when he liked—and that was often. He knew he was an old man, not long for this world. He had still no hope of a life after death, for he would never have committed the ignoble infirmity in his old age of crawling to the feet of a God that he had denied in his prime. But I think he was sweir that all the things he had experienced, and all the knowledge of human nature that he had gathered should die with him, so he passed them on to me over the waters of seventy years. I did not understand a quarter of the things he told me, but this one thing I did understand—that life is both coarse and amusing, that most people are pretentious fools, that fine words butter no beans, and that a wise man observes the follies of the world and goes his own way. Who will say he was wrong? Certainly not I.

It was not only abroad that the Old Man and I spent a lot of time together. Many times when I was coming home from school I would find him sitting on the dyke at the edge of the wood, waiting for me. Gladly I put off the schoolboy and walked round the fields with him, listening solemnly while he told me of the crops he had harvested, of the drains he had cut, of manures he had laid down. He knew every ridge and hollow of his fields, and year by year he taught me to know them so well that I can never forget them. I enjoyed those lessons. The bays of Bohemia meant nothing to me, but the midses of the Home Field were my very life, and I never thought of life apart from them. I did not think then that five years would see us gone from Dungair forever. Perhaps the Old Man feared that his death would be the end of us and made sure that, whatever happened, one of us could never forget the farm. Perhaps—and yet I may be getting sentimental after the event about a man who never fathered a sentimentalism in his life.

Our chief excursion was our drive to church, of which I have already told you. Sometimes my grandmother came too, and Sally would bicycle over and meet us at the door, so that we all went in together. (Uncle Sandy never went near the place because of the strange coincidence that there was always a cow going to calve on Communion Sunday.)

The kirk stood on a little hill surrounded with trees. It was pleasant on a summer morning when the sun shone warmly on the firs, and the bell tolled ponderously in the steeple. But the outside was the best of it. I never did like the inside with its varnished pews, its cold distempered walls, its cheap brass lamps and its scrolls proclaiming heavenly love. It was too much like school and the way that the minister ruled the roost heightened the resemblance. I never felt holy in it, as I often did in the turnip shed at home. On the contrary, due perhaps to the Old Man's teaching, it usually put me in a ribald mood. I found it difficult to be effectively rude in church, and I was limited to coarse imaginings until I fell upon a lovely scheme, lovely in its daring and artistic rightness. It came to me all of a sudden during the children's hymns, when I discovered that the music fitted the words of a certain rude bothy song. I tried over a verse under my breath. Perfect. I hummed the next verse, saying the words in my mind. I joined in modestly with the third verse, and thereafter the church was never dull for me. I got bold in time, and joined rather too loudly in the praise. By good luck only the Old Man heard me. As we sat down at the end of the hymn, he leaned over to me and whispered confidentially—'Gin I wis you, laddie, I widna sing sae lood.' Thereafter I tempered my daring with discretion.

There was one memorable forenoon in church when I sat on the Old Man's hat. Well, I didn't exactly sit on it, but as I was sitting down, didn't I knock in the crown of it. Fortunately it wasn't a tile, but one of those old-fashioned square bowlers. Therefore it was not ruined, but it did look a wreck. This dreadful accident happened at the end of the second hymn, but, as I sat farthest along the pew, I was able to conceal it till the end. Then the solemn moment came when the congregation reached for their hats with a happy sigh, conscious that religion was over for another week. Dungair demanded his hat and I had to pass him the wreck. He gave the hat and me a fierce stare.

'Whatten hell's happened here?'

I managed to stammer out, 'I dinna ken. Somebody's bashed it.'

'Fa wis't.'

I couldn't confess, but my well-trained wits came to my aid. I pointed to Miss Cumming, a weary virgin who shared the pew with us.

'I think it wis maybe her,' I suggested.

'Yea, yea,' he replied. 'We'll see aboot this. Come awa hame.'

We left the church in a horrid silence.

While the Old Man was yoking the grey mare, I had a few words from my grandmother who accused me of having

56

told a lie. If I confessed, she said, the Old Man might let me off, seeing it was Sunday. But I had told my lie and I stuck to it. I knew it was no use dragging in Miss Cumming again, but firmly denied that I had had anything to do with the hat. My grandmother gave up the struggle with a sigh that told me I was in disgrace.

We usually drove home with me standing between the two old people in front, but on this sad occasion the Old Man let down the back seat and perched me on it with one of his grimmest looks, made all the more grim by the disreputable state of his hat. So we drove home in silence, two old people with a load of mischief and misery behind; and I can assure you I was ripe for repentance by the time we reached home, for I was sure the whole countryside must know the reason of my shame. Nobody mentioned the hat during dinner, but though it was stuffed heart, which I adored, I could hardly eat a bite. By the time we rose from the table I could stand it no longer. I sought out my grandmother, flung myself into her lap, confessed and was forgiven at once. But of course that was only the first step.

There was still the Old Man. I must go to him and tell him that I was sorry. I didn't like the idea of that at all and suggested that my grandmother convey my sincerest regrets to him. But no—I must go myself.

'He'll maybe dae something awfu tae me,' I said.

'I wouldna be surprised at that,' she replied. 'Better awa an' get it ower.'

So I went, trembling, to the parlour where I found the Old Man dozing in his big chair beside the fire, with the ruined hat on the table beside him.

'Weel, fut is't?' he asked very grimly.

'The hat—it was me—I did it—an' I'm sorry.'

'Ye young deevil. Come 'ere.'

I went.

'Turn roon.'

I turned.

'Ben' doon.'

I bent.

'Noo than, tak that,' with which he lifted up his great boot and gave me a hoist on the seat of the trousers that landed me in a heap at the other side of the room.

I picked myself up, shaken, but not really hurt.

'Noo than, laddie, awa' get my staff an' we'll gang doon an' hae a look at the stirks in the Moss.'

And that was the end of that.

Incidentally he had a theory that children should not be hit on the head. Nature, he said, had provided an admirable seat for chastisement and who were we to improve on nature? But I doubt if he had any great belief in the punishment of children. He had got plenty of it himself when he was a loon, he said, and damned little good it had done him.

It was about this time that we acquired the Cock o' the North, and life, which was always very full for the Old Man and me during the summer days, became one long adventure after the arrival of that astonishing beast.

There was an old man in our part of the country whose name was Cuddy Manson. He was one of those timeless ancients who seem to have neither beginning nor end, but are a kind of *genii loci*. Cuddy farmed a small croft at the back of the moor, where he raised hens and pigs and did a bit of coping in a small way. He also owned an ass from which he took his name, and a stud boar which had a wide clientele among the small farms of the district. It was a common thing to meet Cuddy driving the boar along the country roads. He used to urge on the beast with growls to which it replied with grunts in the same rough friendly tone; so that there seemed to be some deep bond of friendship between them. Well, Cuddy was passing Dungair one afternoon in his lorry drawn by the ass with a handsome billy goat walking reluctantly behind. He stopped to speak to the Old Man, who took an instant fancy to the goat, maybe because of its cynical air of contempt for the world at large. After an hour's stiff haggling, much enjoyed by both parties, the Old Man bought the goat for two shillings and a dram, I contributing sixpence in consideration of which I received a half share. So the goat became ours.

There never was a wilder spirit trod the earth. He was a perfect devil. He despised the solid ground, preferring roofs for sporting on, and he got his name not because he resembled the Marquis of Huntly, but from his love of perching on the tops of ricks and throwing out a challenge to the four quarters of heaven.

He kept us lively that summer. In the first place he had a tremendous and catholic appetite. He despised grass but adored linen, rope, boots and everything that was never meant to be eaten. If there were clothes on the drying green somebody had to stand guard over them, for he ate a hole out of one of my grandmother's best linen tablecloths and chewed a leg off the Old Man's woollen drawers. He went everywhere and ate everything. One day he got thoroughly entangled in the strawberry bed where he was found eating his way out of a cocoon of herring net. He liked the kitchen too. As soon as Sally turned her back he would jump on the table and finish off a plate of scones in half a minute. When discovered he could disappear like a shot and seldom, if ever, suffered for his depredations.

His sense of fun was peculiar and highly developed. He had a knack of lurking round corners whence he would spring out suddenly on a passer-by and take him sharply in the rear with his horns. Every now and then the peaceful afternoon would be riven with fearful yells, a clatter of tacketty boots, and a raucous derisive 'me-ee-ee-eh.' The Cock o' the North had found another victim. The Old Man, being a bit stiff, was excellent game for the sportive beast and suffered heavy injuries in the first few weeks. However, he was not the man to be put down by any goat. He retaliated in kind. Thus we began a splendid game in which the man and the boy and the goat stalked each other round and round the cornyard. We played the game with spirit, and our senses became so keen that we seemed to develop eyes in the backs of our heads. It was a grand sight to see the Old Man loitering among the ricks with malicious intent, and you realized the glory of revenge when you saw him catch the Cock o' the North an almighty whack across the rump with his blackthorn. It must have stung the old goat a bit, but he took it as all in the fortune of war, and retired only to advance again. I know that stick must have hurt him, because I felt the weight of it myself one unhappy night when I thought to play a trick on the Old Man by advancing stealthily on him from behind. Unfortunately my foot knocked against a dish with a clatter that multiplied itself a hundred times in the stillness of the close. Quick as quick, the Old Man rounded and caught me a rasper across my legs with his stick. It hurt. I let out a yell of pain and surprise that must have echoed across the moor like the howls of the damned, and continued crying diminuendo for the next half-hour. When I had returned to normal, and Sally had blown my nose for me, I gave the Old Man a chance to apologize, but all he said was 'Them that walk by nicht should cairry their muckle feet across their shouthers', and went on spitting apple seeds into the kitchen fire. He did not even offer me a bite of the apple.

The great goat vendetta carried on all through summer and harvest till, I am proud to say, we gradually wore him down. He had an immense advantage in speed and general mobility, but the Old Man had the greater cunning. Besides, there were times when the goat forgot; he had not the Old Man's singleness of mind; and we had an unerring instinct for his off moments. Still, he kept up a magnificent fight until the Old Man discovered the master strategy, which was fireworks. The goat was just plumb scared of squibs and crackers; perhaps the smell of gunpowder reminded him of his father the Devil. We had only to throw a squib at him and he would make one bound for the top of the farthest rick. Yet, though we shook his nerve we could not break his spirit. Like a true Cock o' the North he nailed his flag to the tallest cornstack and prepared to go down butting to the very end.

Alas, his appetites were his undoing. There are some things that even the stomach of a billy goat (and such a billy) cannot withstand. He had eaten most everything that could be eaten, laburnum flowers and all, with no other effect than an occasional asperity of temper. If he had stopped there he might have been flourishing in wickedness to this day. However, he ate a pound of lead paint and that finished him. We found him one morning, dead. It was a strange shock when I saw the beast that had been the very incarnation of energy lying motionless, with his defiant whisker dabbled in the mud. I cried because it was terrifying and strange. The Old Man was very silent all day. He missed that goat because he had been a grand fighter and bore no ill-will. We gave him a splendid funeral at the bottom of the garden, and used the remainder of the lead to paint the barn door in his memory. R.I.P. He was a valiant spirit while he lived, and life was dull without him when he died. So I came to realize the sense of tears in mortal things.

IV
Betsy and William

It must have been quite early in my life when I was taken to town for the first time, but I remember nothing about it, and, curiously enough, the town does not seem to have made any great impression on me. That is very strange, for every detail of many drives home to the farm are as vivid in my memory as if they had happened yesterday.

I have only to shut my eyes to see myself sitting between my grandparents in the gig while the white pony clattered over the bridge in the frosty twilight. I still shiver as I remember the dark road between the moor and the wood and the numbness that used to creep up from my cold toes in their clumsy boots. It was an endless clip-clop journey through the winter evening, but how good it was to climb the last hill and see the light in the parlour window, shining softly among the trees, and what supreme joy were the rositty logs in the kitchen fire, the kippers a luscious bursting brown, and Sally's new-baked scones with honey. These small sensual things remain when all else is forgotten. Perhaps the town was too big for my young mind to comprehend it.

Those early visits to town are associated with a number of old people, the friends of my grandmother. First and foremost among them I place Miss Betsy. She was one of those girls born to poverty and good works and most faithfully she had cultivated her garden. The crop had been a little meagre, for when I knew her she was an inmate of a home for the indigent daughters of the burgesses of the town.

The hospital was charming to my young mind. It was a single-storied place built round three sides of a square with an old sundial in the middle. The two wings, in the cottage style, contained the old ladies' bedrooms, while the centre was a pleasant dining and gossip room surmounted by a bell. Great trees overshadowed the home, the cathedral stood just over the way, and the river murmured to itself forever across the fields. Its garden had the fragrance of late roses; it was faded, elegant and retired; and the passing of the uneventful years seemed to shut it away from the follies of the great world.

It may be deplorable, but it is true, that the atmosphere within the hospital did not correspond with the peace around it. The inmates fought with each other interminably. There was something noble in the way two old women, each with at least one foot in the grave, could carry on a feud with all the impetuosity and more than the cunning of youth. They did not sit down and fold their hands and meditate upon the world to come. They had lived hard lives, for spinsters had no rights sixty years ago; they had become so accustomed to slights that they looked for them everywhere; and their capacity for finding them amounted almost to genius. They were unloved women whose gentler instincts had been thwarted for so long that they had nothing left but hatred. As they lived so much on each other's doorsteps, and as they had been all too willingly forgotten by the rest of the world, they visited their lonely hatred on each other with a malignancy that was truly awesome in women who had such an uncertain hold on life. Miss X and Miss Y enjoyed a vendetta that lasted for many years. Miss X, a small ethereal old lady of seventy, suspected her neighbour of stealing her coal and said so to everyone who would listen. Miss Y, a fat intransigeant sug of a woman, had a tale of two doughnuts that had disappeared when Miss X was about. Miss X retorted that Miss Y was uncleanly in her habits, which may have been true; and Miss Y declared that Miss X held secret orgies of drinking, during which she sang songs of a most abandoned nature. So far the quarrel would be carried on through willing intermediaries, but one day, when the hatred had been worked up to a sufficient intensity, the principals would affront each other in the lobby. The first skirmishes would be confined to general insinuations; then they would advance to particular accusations, and so open up the battle along the whole line of their past lives until their respective families back to Adam would be embroiled. There was no sin or folly known under the moon that the two old women did not hurl at each other. They stormed, they screamed, they raved, they blazed with anger, till it seemed their frail bodies must be destroyed by passion, but neither would yield one step to the other. They were women possessed by all the futility of their barren lives.

Meantime the other old ladies enjoyed the consummation of their tale-bearing. They did not gather round as the vulgar do, but stood behind their open doors, exulting in the din of battle. Matron would soon get to hear the racket—or one of the old ladies would run and tell her, for though they did not wish the fight to be stopped they could never resist a chance of sookin' in with her. However, Matron was a wise woman. She did not stop the fight until the old ladies had exhausted their hatred, then she suddenly intervened, told them what she thought of them, and sent them to their rooms with a few well-chosen remarks. So the noise of battle would die away in a diminuendo of grumbles until the press of hatred would bring the conflict again.

Miss X and Miss Y were not the only ones to fall out. There were as many feuds as there were combinations of eight old women. Sometimes all the feuds would rage together; then the Matron dosed all the old ladies with oil

Sally

and sent them to bed, for they were so old that they had become children again.

Miss Betsy did not take part in the feuds, for she was a good, kind, stupid, Christian woman, utterly without malice or hatred. How she came to be in a family like ours passes all comprehension. She was a tall, thin woman, with a small, bird-like head and a beak of a nose, and as she grew older she became more and more like an aged, melancholy fowl. One can think of her life only with horror. She had always been unlovely and unloved with any fruitful passion. She was a dressmaker to trade and had spent most of her working life as a sewing woman in other people's houses, making new shirts out of old petticoats. Never once had her good name or her immortal soul been in danger. An excursion with the Sunday school or a polite tea party at a friend's house had been her extremest adventures. And yet she was not negative. She had a natural sweetness that increased with the years and a great humility of spirit which saved her from bitterness. She strove to please because it was her greatest joy to give pleasure to others. My grandfather liked her about the house. He liked to shock her modesty with outrageous suggestions about her past, but he also respected her because she was so free from sham; and she adored him because there was splendour in his wickedness that shone resplendent above the little virtues of her own life. The woman had possibilities. It was wonderful to see how she would blossom at a party. After one glass of port she grew quite merry, discovered a store of proper and sometimes amusing stories, and would recite a piece about the death of Little Nell with great pathos. The Old Man enjoyed that recitation. His eyes twinkled more sardonically than ever as he listened. Something in the sight of the spinster who had had so little of life, reciting a sentimental piece about death, appealed to the old sinner who had savoured so much of the one and cared so little for the other. I am sure he often wondered how he came to have such a cousin, and I think he liked to have her about, because the contrast with the rest of the family moved his highly individual sense of humour. But there was more to their friendship than that. Though life, as we say, had denied her many things, she never denied anything in life, never found compensation for her disappointments in a twisted morality. Her attitude was always a humble yea-saying to life—but it was all too seldom that she was called upon to say anything.

She found plenty of scope for her sweetness in keeping the old ladies from killing each other, and, bad as relations were among the aged anchoresses, they would have been ten times worse without her. She was the only one who could persuade Miss X that gin was a dangerous tonic with her morning tea, and she alone could soothe the old lady's nerves when they had been unduly excited by her persecution mania. She was even better than Matron in dealing with the recurrent crises when Miss Z locked herself in the closet to spite her neighbours, and when the old ladies were visited with premonitions of death after a surfeit of pickled gherkins, it was her voice alone that could drive the dark shadow from their beds. Of course she was exploited on every hand, but she loved it. When she died they all said they had lost an angel of mercy—then forgot her within a month, except when they remembered with an almost desperate pleasure that they were still alive while she was dead.

I liked to visit the hospital, but, although it was something different from the farm, it was not very exciting. Now a call on Uncle William in the asylum was a real adventure. Uncle William was not, as you might say, mad, but he enjoyed periods of inspired inconsequence which became very annoying to his family. At such times he was liable to buy a cran of herrings to stock the milldam at his farm, or to preach the gospel in public places, under the impression that he was John Knox, for whom he had an intense hatred in his saner moments. Maybe he was a bit of a nuisance, but it was a pity to shut him away with a lot of real lunatics, for he was an amusing little man with bright blue eyes and a red whisker. He had not the divine simplicity of his Uncle George, who was one of God's naturals. If he had been happier in his married life, and if John Knox had not so vexed his subconscious mind, he would have been an ordinary Scottish farmer. As he was the youngest of a large family, there may have been some weakness in him that could not withstand the strain of life, and his mind took refuge in periodic irresponsibility.

The asylum cured him by feeding him and caring for him and leaving him nothing to worry about. He was a contented little man, did a job of work in the asylum garden, gossiped about old times with any one who had a minute to spare, and so faded easily from a life that had been too much for him to handle on his own.

He was not really an uncle, only a remote cousin, but, as he was in need of affection, my grandmother took him to her heart. We used to call on him (for it was more a friendly call than a visit to an institution) every time we went to town. An attendant showed us into the chapel where patients were allowed to see their visitors. There were usually others beside us, and I suppose it must have been a melancholy business, though I was unable to appreciate anything except that I was among madmen who might suddenly do something desperate. To my disappointment they never did, though my grandmother once saw one who was so influenced by his surroundings that he suddenly rushed into the pulpit and announced that he was the Holy Ghost.

It would be difficult to imagine anything less exciting than that chapel. It was an oblong with high walls lit by those near-Gothic windows peculiar to religious places. The walls were distempered a sickly yellow and the pine forms gave the atmosphere a faint varnishy taste. There we were allowed to see our unfortunate relations.

I remember one middle-aged woman in decent black who used to come regularly into the chapel and take a seat as far away from the rest of us as possible. She sat down, looking straight in front of her, with her hands crossed in her lap. After a few minutes a warder came in leading a man by the arm. The man halted inside the door and looked round with the dazed look peculiar to the place, while the woman half rose and made a gesture to him. Then the attendant led him over to her and left them together. They exchanged a mumbled greeting and sat down, neither together nor apart but like two people on the extreme edge of discomfort. They talked a little in brief, jerked out sentences and briefer replies, followed by long, helpless blanks of silence. The woman produced her little presents—a few sandwiches, a piece of cake, a twist of bogie roll or a paper of snuff which the man received with his first show of interest since he entered the room. Then the infrequent question and answer were resumed until the attendant returned, the man was led away, and the woman went scuffling out of the chapel, looking straight in front of her.

As the man passed us, Uncle William would give him a look of friendly interest and say:

'Aye, that's Jimmie. Puir chap, he has the depressions.'

I am glad to say that Uncle William's depressions were things of the unhappy, and as far as he was concerned, the forgotten past.

He missed the farm in a way.

'There's times I wid like tae be gyaun oot in the gloamin tae hae a look at the kye or tak a daunder doon through the new grass tae the inn, for it's an awfu dry place this; an I'd gie ten poun tae be at the Ploughin' Match jist aince again; but, och, this place isna that bad, though there's some gie queer fowk in't. No, I canna complain. I canna complain.'

He never did complain and he was seldom unhappy, except when he thought of the young cows in milk that had once stood in his byre or when too much gossip about the past aroused the thirst in him. He had a quiet mind. As soon as he got over his first anger at being put away, he settled down to make himself at home in his new world. He knew everything that was going on, and saved up little bits of news for us, so that he could talk a whole hour about the adventures of his companions.

When he had heard all that my grandmother had to tell him about Dungair, and after he had given us his own news, a little silence would come over us during which Uncle William would look out at the window while my grandmother gave me a smile that was almost a wink. Then she took out her presents—the best part of a chicken, a round cake full of cherries, a bag of pandrops and a paper of black twist. When she had laid them on the form, Uncle William returned from his examination of the window and regarded them with a fine mixture of surprise and delight.

'A chuckie—cherry cake—pandrops—an' tabacca—Eh Ann, but ye wis aye that gweed tae me—I'll jist hae a wee taste o' the beastie noo.'

While he enjoyed a cut off the breast, my grandmother would say:

'The tabacca's fae Dungair himsel.'

'Imphm,' said Uncle William through the chicken, 'he's a fine muckle big man yon.' He had always admired his cousin's cavalier attitude to fate.

When he had had his will of the chicken, Uncle William would gather up the remains.

'That'll come in fine for my supper the nicht. An there'll be a leg tae Peter. He's a puir body an he's awfu fond o' chucken, bit there's never naebody comes tae see him nor bring him onything.'

'An' how's Peter behavin?' my grandmother would ask.

'O, he's been a bit better o' late, bit sometimes he gets awfu doon an' then he's neither tae haud nor bin. I've a gie time wi him fin he gets doon, bit then, puir man, he's had a lot tae bear.'

This Peter was a half-wit who had been at school with Uncle William. As a young man he had been just sane enough to be exploited by the farmers of the neighbourhood who gave him board and lodging in return for twice the work of an ordinary man. He became difficult as he grew older, took to violence, and had to be put away. Uncle William found him wandering forever in the wastes of his own mind and adopted him. They became indispensable to each other; Uncle William found something to take care for and he was Peter's one contact with the world of light.

Uncle William and my grandmother were genuinely fond of each other. There is nothing surprising in that, of course, for she was the easiest woman in the world. People felt at home with her immediately. She never pretended to be other than she was. Simply by what she was she lived. She was not humble, but she was so utterly sure of herself that she never felt the need for self-assertion. She was content to be taken as others found her. She never treated Miss Betsy as a poor spinster or Uncle William as a harmless lunatic. They were persons in their own right to her, she treated them like old friends because they were her friends, and they loved her for it. Thus it is that I, extremely sensitive to atmosphere as a child, remember those afternoons in the chapel of the asylum as something rather delightful, when the old people sat and talked peacefully together and the sun came softly through the painted windows.

I am sure that the Old Woman never thought it any disgrace that one of her family should be in an asylum. She had a certain fatalism which would have ascribed it to the Will of God; and, at the same time, she was singularly free from vulgar prejudices. She had a habit of looking at life for herself. There was one occasion when her mother, who was suffering from an attack of indigestion, following a surfeit of partans, made some nasty remarks about Dungair's habit of drinking a droppie too much.

'It's bad an' bad eneough in a man o' his age, maybe,' my grandmother replied, 'but it's nae waur than an auld woman like you eatin partans till her innards gae bad.'

She condemned neither. Maybe they were silly, but 'fowk maun gang their ain gait, and if they dae wrang they'll hae tae pey for't'.

Some people might have asked her what wrong Uncle William had ever done that he had to be put in an asylum. She would just have pointed out that, as he was happier there than he had been outside, it was possible that he had done no wrong, and that the people who had put him there were suffering from the reflected shame of his madness. It certainly was considered shame in those days before the blessed word neurosis had been invented. Fortunately we know a little more about the workings of the mind now, and we are broadening the basis of insanity. It is just as well, for most middle-aged people seem a little mad and none are more dangerous than those who manage to persuade the world that they are perfectly sane.

Miss Betsy and Uncle William have been gathered to their fathers a long time now. Both were incompetent to deal with the life in which they found themselves through no will of their own, and both were despised as failures by people who were unconscious of the mess they were making of their own lives. Yet they were happy in their own way. Miss Betsy was a devout Christian, a little sister of St. Francis who walked with humility in the light of Christ. Uncle William was a pagan whose faith lay all in the earth and the crops and the intercourse of human kind. She died in the hope of a glorious resurrection; he because he could not live any longer. Though they may seem to have been as wide apart as Heaven and earth, they shared an essential simplicity. I am glad I knew them, because Miss Betsy taught me not to despise the poor spinster who cannot marry and will not burn, and Uncle William gave me a taste for amiable lunatics—so long as they are restrained from the direction of public affairs.

And now is the time to introduce some more of the remarkable characters who walked the paths of this world when I was a boy. Let me begin with Besom Jeanie. Perhaps she had another name entered in the registers somewhere, but Besom Jeanie was all I ever knew her by. She was a little woman whose gnarled brown features, deep-set, glittering eyes and heathy smell gave you the impression of antiquity rather than of mere age. She was old, maybe sixty, maybe eighty, maybe a hundred, who knows, but she had a man whom she called her husband and who might have been her son, who was not a day more than forty. Jeanie used to travel our part of the country with a basket of scrubbers made out of heather which she exchanged for money or kind. She visited Dungair perhaps once a month about dinner-time, sat down at the kitchen fire without invitation and waited for the bounty which never disappointed her. She had a phenomenal appetite; maybe she was like a camel, able to eat enough at one visit to sustain her till the next; anyway she ate enough for three men, and filled her basket with as much again. A dozen oatcakes, a roll of butter, a twist of tea and a paper of sugar—these were her regular allowance, with a dozen mealy puddings, or a bone for soup, at an odd time. The Old Man liked her, for she knew all the gossip of four parishes and he could always spend an hour with her at the fireside. He could not abide her man whom he called a sneaking, scrounging, good-for-nothing (but in quite other and far more vigorous terms) and he took every opportunity of playing tricks on him. There was one high occasion when the man, being on the scrounge, asked for some onions that were drying in the garden. Dungair told him to help himself. Jock took all the onions his pockets would hold and departed with profuse blessings on the Old Man's unhallowed head. That night Jeanie and Jock were camping on the moor. They cooked the onions and had a grand blow-out on them. Next morning a shepherd found them half dead with colic. No wonder. The onions were daffodil bulbs, a fact of which the Old Man was perfectly aware. It was months before Jock came near Dungair again.

Now and then we had a visit from a more picturesque and, to my mind, far more interesting character. This was the Buccaneer, a huge hearty man with a wooden leg and a crutch, who used to pursue many strange agencies under the guise of selling bootlaces. Dungair called him the Buccaneer because of his wooden leg, and the name was really appropriate, for he had a roving air, a ready but dangerous smile, and a speaking eye. His mouth was large, his teeth irregular but white, and he had one of those laughs that shake the ceiling. He fascinated me because he was so completely master of his world and free from all the humiliating disabilities that hedge the lives of little boys. He had wit. He could tell a story, hundreds of stories. He told me at least a dozen different versions of the losing of his leg and I believed them all. He was an artist, above criticism. I was sorry for him too because he said his wife had run away with another man. She had been madly fond of ducks' legs and green peas, he said, but after he lost his leg he could not afford to give her any. A man came along who was in the poultry world and lured her away with promises of unending duck. She yielded to temptation and went. As the Buccaneer had a wooden leg he couldn't run after her—so there he was a lorn man with only one good leg to his name. My heart used to grue at the hearing of the story—but what good could I do? I was only a poor boy and I never felt more bitterly the ineffectiveness of youth.

However, the wooden leg did not incommode the Buccaneer unduly. He was a roistering fellow on Saturday nights and quite often started a fight in the low quarter of the town. Although you might not have expected it, he was a doughty scrapper. Once let him get his back to the wall and he could take on the whole world, swearing horribly like a Brigade of Guards, and laying about him with the crutch, like one of the heroes of old. He had broken more heads than any other two men in Scotland, and they with all their legs complete. He was a splendid man, with a clear-springing vitality that never yielded to adverse fortune. He was one of my major heroes.

I must not forget Mr. Thomas. He was a dear old gentleman of a class which has completely disappeared. He was the last of the real packmen. The present state of the world has driven many men to hawking goods from door to door, but they are not packmen in the true sense of the word, not packmen of the order of Mr. Thomas. He was a travelling merchant who, though he carried his stock-in-trade upon his back, had all the pride of a city merchant. He was a square, stocky old gentleman, always very neatly dressed, with a pink face framed in well-trimmed side whiskers. He carried his goods in a square box slung over his shoulders and thus he would tramp the country, trafficking with the farm people and villagers, and return to town in a month to replenish his stock. His goods were various—cheap watches, cheap jewellery, studs, pins, needles, lace, spectacles, all kinds of odds and ends for young men and maidens. He was a walking Woolworth, but he did not charge Woolworth prices. He took what he could get and if he took a profit of five hundred per cent. he earned it by the many dusty miles he tramped with the square box on his back. Then he bought horsehair from the farm boys, human hair from the girls, and rabbit skins and moleskins from sportsmen like Dungair. He didn't like to pay out good money of course, preferring to give goods in exchange, and it was sound business to trade a tiepin value a penny for a shilling's worth of horsehair. He did a big trade in spectacles, for, if he could not sell you a pair that were just right, he could at least give you something to help your failing sight. So it would be a mistake to call him a hawker. He was a very comfortable business man who had a house in town and money in the bank. I confess he was not very interesting in himself, but he was interesting as one of the last of the packmen who were once an important factor in the world's trade. It is perhaps as well that they have disappeared, for they made ridiculous profits out of the ignorant country people—and yet they were a picturesque company, delightful gossips, altogether a valued amenity in the rural life that even then was passing away.

V

The Steading

When I was a tiny little boy my whole world had been enclosed in the house and garden at Dungair, but now that I was going to school I entered into enjoyment of the steading and the fields as well. I even began to despise the house a little because, after all, it was only a place for women to work and men to sleep in, but the steading was altogether of a larger world.

As I have already told you, the farm buildings enclosed a square court which acted as the clearing station of our daily activities. If you faced north from the kitchen door you had the stable on your right, the barn in front of you, and the byre on your left, making three sides of the square, while the house made the fourth with a ten-yards carriage space between it and the gables of the byre and stable.

I have mentioned that the court was usually littered with agricultural all-sorts, but it may amuse you if I tell you a few of its more remarkable features.

In the first place there was the great rainwater barrel that stood outside the kitchen door. It must have been the father and mother of all barrels, for it was about seven feet high and round-bellied as a bench of magistrates. I remember the day it arrived at Dungair and I remember the rich winey smell that hung about it for years and fined away gradually as the rainwater soaked into the grain of the wood, once perfumed with so much good wine. I don't know what its history may have been; perhaps it had once held a noble vintage, or a solera hallowed by centuries; but I am stricken with shame that I should have been a party to the degradation of such a vessel to the storing of rainwater, as one might use a Caesar's bones to keep the wind away. Time passed, and its smell gained in potency what it lost in grace, as I discovered one afternoon when I chose to jeer at Uncle Sandy. He happened to be in a genial mood that day which encouraged me to throw a piece of decaying turnip at him. The turnip hit him between his hair and his collar with a most unpleasant 'zug' while I stood thrilled by my audacity and horrified at its success. I didn't stand there long. Brushing pieces of turnip from his neck, Uncle Sandy pounced on me, lifted me by the slack of my pants, swung me high over his head and dumped me into the water-barrel with a mighty splash. I was fortunate. The splash was caused not by the depth of the water but by my landing in it bottom-first, which caused a considerable displacement. When I got over the first shock, I scrambled to my feet in eighteen inches of dirty green water and let out a cry of rage.

'Are ye droont yet?' asked Uncle Sandy, looking over the edge.

'Aye, I'm droont; I'm droont; I'm droont,' I bawled.

'Aw weel, droon awa,' he replied, and disappeared.

I heard his feet go clumping over the cobbles towards the cornyard. A vast silence filled the court.

But not for long. I raised a cry. No one came. I cried again. Still no reply. Then I roared as only I could roar and banged on the slimy walls of the barrel which multiplied my yells with hollow and redundant sounds, deafened me with the multiplication of my own riot, deafened then frightened me, till I suffered a moment of panic when I thought the concave walls of the world were about to close on me. Then a hen that always nested in the old pigsty across the court rose from the straw and announced her daily miracle in a paean of self-advertisement. That familiar sound restored my self-possession. I wiped my tears, blew my nose, and resigned myself to the inevitable. I was really in no great discomfort, for it was a hot July day, the water was cool if stinking, and I had nothing to do, which has always been my favourite occupation. So I made myself as comfortable as I could and began telling myself a fairy tale in which Uncle Sandy figured as the wicked ogre. It was a good story and I thought up so many original forms of torture for him that by the time he returned to rescue me I felt quite sorry for him and forgave him everything. I suppose the experience should have done something terrible to my psyche, given me nightmares and afflicted me with a persecution complex, but we had tougher psyches then, and I took no harm from the experience at all. Why should I have? It was only Uncle Sandy's idea of fun and where could you have cleaner fun (theoretically) than in a water barrel.

The stone cheese-press stood beside the water-barrel. It was a small but powerful bit of work consisting of two upright slabs of stone capped by an ornamental transverse slab. A great block of hewn stone, weighing about two hundredweight, hung from the transverse slab by an iron bar which passed through a hole in the centre of the slab. The iron bar was worm-screwed, so that the block could be moved by a lever on the top of the slab. My grandmother used to make excellent cheese in this press, crumbly, blue-veined country cheese that took you by the throat and shook you. We never used any other at Dungair, and it was a long time before I could believe that the soapy stuff they sell in towns was really cheese at all. Ours was honest native cheese; it had the keen edge and tang of the north-east in it; and the only thing I have ever found to equal it in sound local flavour, or, you might say, pride of race, is an occasional Wensleydale and a rare Stilton from the Stewartry. My grandmother made her cheeses when milk was most plentiful, then stored them away in the coolest, darkest corner of the dairy where the secret processes of gestation went on until they came to the table, blue-veined as an aristocrat and frisky as lambs in spring. When you ate a whang of that cheese you knew you had something inside you; it made a perfect lunch with oatcakes, new butter and a jug of harvest ale; and there was nothing better for tamping down a hearty meal, because, as the saying is, it digested everything but itself.

Between the cheese-press and the water-barrel there stood the kennel, inhabited from time immemorial by the long line of our farm dogs. Perhaps the fact that all its tenants had been mongrels made it congenial to me. Anyway it was my constant retreat when the world became too much for me. The Old Woman and Sally were usually sympathetic when my inquiring mind had brought me disaster, but there were times when even they thought I deserved to stew in my own juice. I retired to the kennel for the stewing and there I used to crouch among the mouldy straw and bones, sobbing out the desolation in my heart, abandoned by the world and alone. A mongrel I possessed for a short time, a flea-bitten and much persecuted hound, used to join me in the kennel and lick away the tears as they ran down my fat cheeks. His air of fellow-feeling touched me profoundly. I put my arms round him and, as I hugged him, my loneliness passed away and I fell asleep till suppertime when I forgot my sorrows in the pleasures of the table. I suppose that hound should have made me a dog-lover, but I am happy to say that he once bit me in the leg when I was just testing how firmly his tail was attached to his body. His base ingratitude completely cured me of any tendency I might have had towards idolizing dogs, which is one of the major human lunacies.

The stable was something almost sacred when I was a little boy. As it was private to men and horses, and as no woman dared enter except on sufferance, it summed up all that was desirable in manhood. I could imagine nothing more desirable than to become one of the freemasonry of

the horse, to have the gift of the horseman's word, to understand the esoteric country jokes and to sit on the cornkist
like a real man. It was not an ignoble ambition; until I was
twelve I had no other; and I would have attained it had not
the war brought money to Dungair, made higher education
possible for me and so completely changed our scheme of
things. As it was, I learned a great deal in the stable that is
not in the curriculum of any school.

I need not describe the stable; it was the same as ten
thousand others in Scotland; and had the same smells of
ammonia, leather, and horses. Its chief furnishing was the
cornkist that stood in the window recess, with its lid polished
black by harness oil and the seats of corduroy breeks. Some
fine talk went on around the kist while the boys smoked and
spat and took the countryside through hand. They may have
had just three topics of conversation—women, men, and
horses, in that order—but what they lacked in generality
of interest they made up in particularity of knowledge and
an Elizabethan directness of expression.

The cornkist had an important place in rural tradition,
for the old country songs are so closely associated with it
as to be known as cornkisters. You see, the boys sat on the
top of the kist and kept time to their songs by thumping
their tacketty boots against the front of it. If you had been
passing the stable door of a summer evening when the sun
was going down and the west wind was blowing softly over
the young corn, you would have heard the mournful words
of 'The Dying Ploughboy' sung with tremendous pathos by
a deep baritone, while half a dozen heels beat time against
the kist, like distant thunder, or the insistent beat of fate's
winged chariot hurrying near.

> It's jist *thump* a week *thump* ago *thump* the morn
> *thump*
> Sin I *thump* wis weel *thump* an hairst—*thump* in
> corn *thump;*

> Bit some-(*thump*)thing in *thump* my heart *thump*
>   gaed wrang *thump*,
> *A* ves-(*thump*)sel burst *thump;* the bluid *thump*
>   ootsprang.

And so on through as many verses as the singer could remember or improvise. The cornkisters are not all as mournful as 'The Dying Ploughboy'. They express the whole life of the farm servant—the ecstasies of love, the miseries of marriage, the meanness of masters, the greed of mistresses, and above all the pride in the plough. Some of them reach the heights of Rabelaisian humour; others sound great depths of bathos; but many of them are true and lovely songs. The finest of them all used to be sung by a foreman we once had who combined ruthlessness in pursuit of his pleasures with a romantic taste in song. You did not know what lyrical feeling was until you had heard him sing:

> Braid up your gowden hair,
> Pretty Peggy my dear;
> Braid up your gowden hair,
> Pretty Peggy-o;
> Oor Captain's name was Shaw,
> But alas he wede awa;
> He died for the love of a lady-o.

By the exquisite melancholy in his voice you knew that he felt deeply the romantic fate of Captain Shaw. Perhaps he thought: 'There, but for the grace of God, go I'; but I am sure that, if there had been any dying to be done for love, he would have left his ladies to do it. The truest of the cornkisters is 'Drumdelgie', which tells of the hard service at 'a fairmtoon up in Cairney'. After a recital of all the miseries suffered there, the singer concludes with the excellent and typical valediction:

> So fare ye weel, Drumdelgie,
> I'll bid ye a' adieu

Singing 'cornkisters'

An' leave ye as I found ye—
A maist onceevil crew.

After the war it seemed as if these fine old songs, the ballad minstrelsy of the north-east, would be forgotten, but Mr. Willie Kemp and the gramophone arrived in time. Mr. Kemp has recorded the songs with the true artlessness and the right touch of stormy weather in the voice which characterize the native woodnotes of our rather windy acres. The cornkisters were my introduction to music and they formed my taste so soundly that I have always been able to reject drawing-room ballads, vocal waltz refrains and torch songs at the first hearing. We were taught in school to sing 'Ye mariners of England', so that a wholly agricultural community might show its devotion to the sea. It would have been more to the point if we had been taught 'The Muckin o' Geordie's Byre', for at least we would have understood and enjoyed its humour. But I am afraid it was not proper enough to be used in educating children for a life which is full of improprieties.

The hayloft above the stable always fascinated me with its floor which had been polished smooth as a mirror by a hundred hay harvests. It had a door in the gable end through which the hay was forked from a long cart and packed in tightly to the roof. The loft was a grand place for sleeping in. Once deep in the hay, you felt secure from the world, and I spent many afternoons there when I should have been helping in the fields, taking care not to descend until everybody had forgotten that I had disappeared, for I was a master in the art of returning as if I had never been away.

Old Willie, an ancient whom Dungair employed because he would believe anything he was told, liked the hayloft as much as I did. He had no objection to a little digging in the garden, but when there were turnips to be pulled he would just disappear for a few hours among the hay. By and by the Old Man, having missed him, would go into the stable and cry:

'Wullie, Wullie, ye lazy aul deevil, whaur are ye?'

At the third time of crying Wullie would leave his comfortable bed among the hay and come down the ladder into the stable. As soon as the first foot appeared through the hole in the floor, Dungair would begin a comminatory oration which grew in eloquence as the legs descended cautiously, worked itself into an artistic fury over the tattered back, and broke like a star-shell over the apologetic face which Old Wullie turned to his master. Wullie heard him out meekly then at the first opportunity he would say:

'I wis jist deeing a jobbie for the Mistress,' take off his bonnet, and produce an egg from it.

The two old gentlemen looked solemnly at the little white egg in the big brown hand—Wullie with a touch of anxiety and the Old Man with growing scorn.

'It's a fine bit eggie,' the ancient would venture to remark.

'Tchach,' Dungair would reply. 'Three hours by the clock tae lay a thing like that. Ye'll need tae dae better than that, man. Hirstle yont.'

Having replaced the egg in his bonnet, Wullie would hirstle yont across the court on his enormous splay feet and the Old Man would give me a dry look, saying, 'If ye're gyaun tae tell a lee, tell a big ane.'

Wullie worked the egg-alibi for years, and I must say he was particularly good at finding the secret places in which the hens chose to make their nests; but then he spent a lot of time in thinking up just such places where he might pass his declining years in sleep. You couldn't blame him, of course, for he never asked for wages and I am sure he never got any.

The barn gave me the horrors. I used to have nightmares about it in which a giant without a head tried to feed me into the mill. First of all there was a long chase while I ran about, like a helpless beetle, among his feet, it seemed for hours; then he held me round the middle, pinning my arms to my sides with his huge claws, and pushed me towards the drum. My eyes could not shut, nor could I turn my head away, but was forced to watch the horrid spikes of the drum flash round at a monstrous pace, while the sound of the machine rose to an intolerable hum. I could not struggle, for I was held so tightly; I could not scream, for my tongue had died in my throat; I could only look at the flashing teeth into which I was being inexorably carried. Then, just as I seemed lost forever, I awoke. The moment of wakening was hardly less terrible than the dream, but as my heart returned to its normal beating, my relief at finding myself in bed was so great that I fell asleep till morning.

There must have been something sinister about that barn. Indeed, parts of it were eerie enough. For instance, you could creep into the threshing mill, disused for many years, to be smothered by cobwebs and scared by showers of mouldy corn that fell on you unexpectedly. Or you could creep along the driving shaft to the mill-course outside where the horses supplied the motive power by pulling at a beam. That was a tremendous adventure to a fat and imaginative child, for the tunnel was thirty feet long, dark, damp and narrow. I might have stuck in the middle and died unknown to the world outside, while the rats, of which there were legions, ate me to the bare bones. My courage never lasted beyond the first few yards, but, as I was too fat to turn in the narrow space, I had to go on to the other end. I was always so relieved to see the light again that I did not worry about the nettles that grew in the pit into which the tunnel opened.

Besides the mill, the barn contained a bruising machine that prepared the oats for the horses, a winnowing fan that created a beautiful dusty draught, and a chaff-cutter in which a cattleman once lost half an arm. Its most notable tenants, however, were the rats. The walls and floors were undermined with their runs and their holes in the corn room had been mended so often with tin that there was hardly an inch of wood or plaster to be seen. Everything possible had been done at one time or other, but the plague of rats remained undiminished. Systematic poisoning had only a temporary effect, and usually meant the death of a few dogs and cats as well. Shooting was the only satisfactory way of dealing with them, and we used to have some excellent sport in the summer-time. The best plan was to wait till the quiet of the evening when you put down a handful of hens' meat outside the barn-door and hid yourself in the pigsty with a shotgun. There would be twenty rats fighting over the hens' meat within five minutes. You took careful aim and loosed off both barrels into the middle of them from a distance of fifteen yards. Those that were not blown to pieces got something to remember you by for a long time.

The byre was the gentlest and sweetest place about the steading, just as cows are the gentlest and sweetest beasts about a farm. I have wondered for a long time why men chose to make the dog their closest friend, for the dog is neither very beautiful in himself nor is he particularly virtuous. But the cow—there is a thing of beauty. She is simple, sensuous and impassioned, like a poem, and she has the Christian virtues as well. She embodies the maternal impulse magnified to the nth degree. She asks but to give and give and give again. In return for a few turnips and a handful of straw she gives streams of lovely milk, rich and health-giving, the very elixir of life. The dog is faithful, but the cow is no less, nor will she ever bite the hand that feeds her. The dog has soft brown eyes; but so has the cow; and the eyes of a cow shine with a soft enraptured light, like moons in a misty sky. The dog is noisy and demonstrative; the cow has an infinite capacity for silence. The dog is a complete extravert and snaps at every passing fly; the cow lives in a deep philosophic calm; she has long thoughts and keeps them to herself. To sum up: the cow is a contented beast and spreads her contentment about her; she is beautiful in her life and in death she can be yet more beautiful when properly divided. She is kind and gentle; in short, she is a perfect female friend. I have always loved cows so much that I would become a farmer if I could afford it. I can imagine nothing that would more gladden my heart than a byre of forty feeding as one.

Forty was the number that filled the big byre at Dungair. They were not all cows in my time, for the Old Man had given up the dairying, but there was always a byreful of feeders in the winter-time. It was my delight on the haary gloamings to help the cattleman with the nowt. A few dusty oil lamps cast a weak shadowy light on the level backs of the cattle, the byre was sweet and warm with their breath, and the atmosphere was kind, domestic. Jimmie, the cattleman, liked his beasts. He was never rough with them and swore at them as if they had been his own children. Thus they were never afraid and never excited. I played about among their feet in perfect safety. It was good when they were all feeding, for they made a grand chumping noise as they bit into the juicy turnips and sometimes let out sighs of deep contentment. But it was even better when the Old Man and I went the rounds at night. Then they would all be lying in their stalls, two and two, ruminating steadily. They made no sound but the low grinding of their jaws, or the rattle of a chain as one raised her head and turned to show two soft eyes shining in the light of our lantern. It was then I knew them as the loveliest of the beasts, for the horse and the dog have taken on something of a human temper, but the owsen have not risen above the earth. They have the bounty and the kindliness of the magna mater. And, as we stepped quietly through the byre, we felt a close community with the generous heart of nature.

A doocot hung at the gable end of the byre. It had been untenanted for years, but a sudden whim of the Old Man to have pigeon pie on tap made it a source of pleasure and of a very characteristic type of altruism. By some means, fair or foul, but most likely foul, we raised enough capital to stock the cot with pigeons. They were delightful little persons as they strutted about the court picking up corn. I made a fantasy about them in which they were princesses charmed by a wicked fairy, and I have no doubt the Old Man made fantasies connected with pigeon pie. They were a grand investment, for they were not only beautiful but they also increased at a gratifying rate. The Old Man and I could never have done congratulating ourselves. We looked forward to the time when we would be the biggest pigeon owners in Scotland. We even thought there might be money in the business if we could train them to feed on our neighbours' corn. It was just then, when things were going so well, that my grandmother, with the touch of malice which even the best of women have, thought up a scheme for the good of our souls. Whenever any small boys came to visit Dungair, she would suggest that we give them a pair of doos home with them. The first time that happened the Old Man and I rose up in complete refusal. It was the most ridiculous idea that we had ever heard. But my grandmother insisted in her quiet way, the rest of the company agreed that it would be a pretty gesture and the small boy began to act as if the doos were his already. Some fool of a woman then said how nice it was to see such generosity in the young, and the pernicious small boy decided what pair he would have. The Old Man and I looked at each other in bitterness of heart. We had been manoeuvred into a false position. We could only accede to it as gracefully as we could. That brat of a child took our two best pigeons, but I am happy to say that I was able to push him into the midden before he left. The Old Man and I were in no very good humour when my grandmother told us we ought to be all of a glow because of our good deed. Dungair's reply was unprintable and I, though I had not the words, had much the same emotion. Yet the glow came when we least expected it, for the pigeons returned overnight.

'Aye, aye,' the Old Man said when I told him of the miracle, 'cast your bread upo' the waters ...'

My grandmother never again had to suggest to us that we give away our pigeons to nasty little boys. We took an intense but secret delight in such generosity, and, if we gave away those doos once, we gave them away twenty times. They always came home by morning. It was a good game and a good moneymaker too, for some fathers were so pleased with my generosity that they tipped me a shilling, or even half-a-crown. Dungair made out a claim to a half-share of the tips, which I denied *in toto* and with heat, but I saw he had some justice on his side, so I bought him an ounce of Bogie Roll from the travelling grocer. That surprised him so much that he gave me a shilling, a thing he had never done before and never did again. Thus did I discover the value of a well-considered generosity.

I have told you that the west side of the garden was sheltered by the back of the cart-shed. The open front looked out across the fields to the west and caught the evening sun that used to come flooding down the valley at mid-summer like a river out of heaven. The cart-shed, however, was very much of the earth, for, although it did hold half a dozen carts besides odds and ends of broken tools, it was the favourite promenade of the hens. It had an earthen floor of beaten clay, but so much mud from the cart wheels, so many dead leaves, such quantities of this and that had accumulated throughout the decades that it was carpeted with a fine dross in which the hens took baths and even made their nests in the more secluded corners. Particularly it was their refuge on rainy days when mud lay inches deep all over the farmyard. Then they would stand under the arches on one leg and watch the drips falling from the leaky gutters overhead with their heads cocked wisely, now on one side, now on the other. Nothing quite equals in profundity the look on the face of a hen who is watching nothing in particular, for that is exactly what her mind is able to understand. I have a very low opinion of the hen.

That is very ungrateful of me, because the hens were my earliest and by far my most appreciative audience. You see, when I was six or seven, I fell in love with preaching and discovered an ambition to be a minister, It was not that I had found grace—I never have, though I had a narrow escape after my first unhappy love affair when I was sixteen—but I realized how glorious it must be to speak to masses of people who have no chance of speaking back. I made up my mind that I would find an audience. Well, I knew better than try it on Dungair or the men in the stable. That left only my grandmother and Sally, but they were only two and I wished a full congregation. While I was casting about for my heart's desire, I chanced to look into the cart-shed on a rainy afternoon and see a hundred fowls, a score of ducks, and half-a-dozen turkeys, standing about on one leg with their usual air of patient attention. There was my congregation. They were obviously waiting for something; perhaps they were waiting for me. They did not have long to wait. I ran into the house, got my Bible and hymn-book, and ran out to the shed again, where my congregation were still waiting on the other leg. I climbed into the seat of a disused reaper, draped a cornsack over my shoulders for a robe, gave out a hymn and sang it with all the fervour of my little pipe. Then I prayed over them, read a chapter of the Bible and dismissed them with an extempore benediction. No preacher could have wished for a better congregation. If they showed no evangelical fervour, they paid me profound attention, cocking their heads from side to side and

Preaching to the poultry

regarding me with round, brilliant eyes. Occasionally they gave little nods, as if approving Holy Writ; and sometimes contented clucks, as if assured of my orthodoxy. Only the turkeys, the stupidest of all created things, failed me in due solemnity. They looked at each other anxiously, hopped from one foot to the other, and even broke into an absurd gabbling at my finest passages. When they did listen to me, they had the air of red-faced churchwardens, whose minds were on the offering. But the hens—they were a model to any congregation, attentive little women whose bright eyes never dimmed with slumber. How well they pretended to enjoy the lineage of the Kings of Judah. I often conducted those cart-shed conventicles; indeed, during one long wet spell I read all of Genesis and most of Exodus to the hens. Of course my game was discovered and brought me a great deal of ridicule, but I welcomed it, for I had been told it was the portion of the Saints. My grandmother encouraged me. She thought it was maybe a sign of grace and that I might be a real minister some day. She even spoke about it to the Old Man when we were sitting round the parlour fire one evening.

'He'll maybe hae a great big Kirk o his ain yet,' she said.

'Imphm,' said the Old Man, 'wi a family like this ahin' him? Then God hae mercy on his congregation.'

He was a remarkably farseeing old gentleman.

The evangelical fit did not outlast the winter. Of course it has returned to me several times, bringing with it a passionate desire to save the world. Fortunately it has never lasted long enough to make me minister of a Church— fortunately for the Church, I mean, because I am sure that I would have disrupted it in six months and would likely have spent my life leading small bodies of dissent into a spiritual wilderness in the name of truth.

The poultry at Dungair had thus the chance to find grace, but there was a community on the farm who had little chance of grace and desired less. I mean the ploughmen. They lived in the bothy, a one-roomed house across the yard from the cart-shed. Unlike most of its kind this bothy was pleasant enough, for its back window looked west to the valley and the hills. The furniture was of the simplest—two big double beds filled with chaff, a wide open fireplace for burning peats, a tin basin to wash in and a roller towel behind the door, and a spotty mirror in one of the windows. Pitch-pine walls and a cement floor looked almost as cold as they were, but they had the necessary merit of being clean. Cold and clean but never a home, you might have said if you had seen it at Whitsunday weekend when the old men had left and the new ones were not yet home, and after Sally had spent a day in scrubbing it out with soap, soda and ammonia. If you had seen it a fortnight later, when the new boys had moved in, you would have found it neither cold nor clean nor any more like home. The farm servant in those days—and I suppose there has been little change—had only two possessions—a kist and a bicycle. So, if you had looked into the bothy, you would have seen the kists set out against the walls and the bicycles in a recess at the foot of the bed. Sunday suits, shirts and long woollen drawers hung from nails on either side of the windows, each man having a bit of the wall for wardrobe. Boots, ranging from stylish browns for Sunday to great tacketty boots all glaur and dung, huddled beneath the beds where they had been lightly thrown off their owners' feet. A strange collection of things littered every shelf—bits and pieces mostly broken, collar studs, screw nails, jews' harps, cogs, flints, gas burners, ball bearings, old knives, corkscrews, cartridges, bicycle clips—everything for which you might find a use if you kept it seven years. Anything they really valued they kept locked in their kists. But you must not think that they left only rubbish about or that they made no attempt at decoration. Most of them had photographs nailed up beside their beds—photographs of relations, very self-conscious in Sunday blacks or white elbow-length cotton gloves; photographs of horses in gala trim on the way to a show or a ploughing match; or photographs of ample ladies in a state of frilly *déshabillé* who must have been left over from the Gay Nineties. These last were real art and treasured as such. Sometimes the reverent owners enclosed them in wreaths of strawplaiting, such as they used in horses' tails, and I suppose the Gaiety Girls must have looked strangely bucolic enclosed in 'long and short' and peeping coyly from under a head of corn. Still I thought they were lovely ladies and so, I am sure, did their owners.

The etiquette of the bothy and stable was equalled in rigidity only by that of the Court of Louis XIV. Each man had his place and was taught to keep it. For the second horseman to have gone in to supper before the first horseman would have created as much indignation as an infringement of precedence at Versailles. The foreman was always the first to wash his face in the bothy at night; it was he who wound the alarm clock and set it for the morning, and so on. The order of seniority was as strictly observed between the second horseman and the third, while the halflin always got the tarry end of the stick. The cattleman's status was indeterminate; I rather think he was on his own; but, as he tended cows while the others worked that noble beast the horse, he was always regarded as inferior, whether he admitted it or not. But the foreman had pride of place in everything. He slept at the front of the first bed—that is, nearest the fire; he sat at the top of the table in the kitchen; he worked the best pair of horses; and he had the right to make the first pass at the kitchen maid. His character had a considerable influence on the work of the farm; if he was a good-tempered fellow he kept the others sweet, and if he could set a fast pace at the hoeing he could save pounds for his master.

The ploughmen usually rose at five in summer and half-past five in winter. They went to the stable at once, fed their horses and then came into the kitchen for breakfast. Yoking time was six from March onwards, and from daylight in winter. They stopped for dinner at eleven, then yoked at one and lowsed for the day at six, or dark. In harvest they might work on till ten or eleven, if the dew did not fall heavily, and I remember two autumns at least when the binder worked till midnight under the great red harvest moon. As soon as they had fed their horses they came into the kitchen for supper and in winter used to remain at the fireside till nine o'clock, telling stories or playing cards, when they looked to their horses and retired to the bothy for the night.

Sometimes a few of the boys from the neighbouring farms came to see them. We would all go out to the bothy then and lie on the beds while somebody played the melodion and we sang the traditional songs of the countryside with variations to suit our mood. I enjoyed those parties. The peat fire glowed with an intense smouldering heat; the paraffin lamps burned dimly, for there was always a black comet on the glass; the bothy was warm and smelled of hard soap and human kind. I lay on the foreman's half of the bed, three parts asleep, and listened to the melodion, or joined in leisurely songs where the beat was held up interminably for romantic effects. But, no matter how hard I tried to keep awake, the sleep overcame me, and I sank down into the deeps of bliss, troubled only by the gales of laughter that saluted some hardy tale. Then the foreman carried me into the house—and morning came in a long moment.

And now I think I have told you enough about the steading to give you an idea of the life that had gone to create it; and I will make an end of this chapter by trying to assemble the purely physical impressions so as to give you a picture of Dungair. Imagine then the main buildings and the house making a square around the close, backed by the corn-yard and fronted by the garden. Imagine the cart-shed on the west side of the garden facing out across another yard to the bothy and the pump. And imagine a treble line of beeches planted along the west in the field behind the bothy, planted by John and grown into their vast maturity after seventy years. Then imagine eleven fields grouped around the houses, one hundred and eighty-six gently brae-set acres, sloping away to the south and west. And then imagine woods and fields stretching far and far along the valley into the blue mists of a summer afternoon, until the hills joined hands in one coned summit across the horizon, on the very marches of infinity. In that spare but not unlovely land I of the sixth generation grew up to be a farmer's boy.

VI
1914

The summer of 1914 was memorable for two things—my grandfather got clear of his trustees for the third time and the Great War broke out. The first of these events had an incalculable influence on my own life, but the second was perhaps even more important, for it completely changed the old world in which I had been reared. It may be only sentiment on my part, but I look back to the summer of 1914 as the sunset of a world of greater ease and personality, when paganism had an exquisite bloom of decay and old customs flowered in an Indian summer of tradition.

Curiously enough, this is the first year that I remember as a whole. My earlier years are no more than isolated impressions against the background of my grandparents' love for me. In this year I seemed to become conscious of my whole world.

Let me begin with the Christmas treat at the school. It was always held on the day we broke up for the holidays. During a day or two before, good-will and irresponsibility stirred like tender green shoots in the rather frozen surface of academic life, to burst into gaudy flower on the last day. We were supposed to work all morning, but how could we in rooms which had been decorated overnight with amazing streamers and the like. Besides, farmers kept coming and going all morning with boxes of cakes and apples, tea-urns, and penny crackers which they stacked in the porch with

roars of laughter as they chaffed Miss Grey in a most sacri-
legious fashion. Miss Grey hopped about like a hen on a het
girdle, half in the room, half out of it, and wholly distracted.
Her grey hair escaped from its iron discipline and waved in
the wind created by her own impetuous motion. Her eyes
flashed all ways at once. Her voice rose higher and higher,
dominating, imperious. I have never seen anything more
like Britannia riding the whirlwind and directing the storm.
By noon she had forgotten us altogether, except when she
required some of the bigger boys to carry boxes into the
kitchen. It was not unmixed pleasure to work under her
in such a mood, for if her orders were not obeyed to the
sixteenth of an inch, her wrath was instant and devastat-
ing. Willie, a complete bonehead who is now a cottar man
with half a dozen children, did not understand that when
she said the cups must go in the corner she meant in the
corner. Near enough was near enough for him, but not
nearly near enough for her. She jumped at him like a pan-
ther and hit him with the first thing that came to her hand.
It was a carton of cream cakes (for the ladies who were to
assist with the tea) and burst with a fine glaury effect on
his round turnip head. He stood there in front of the class
with whipped cream and bits of meringue sticking all over
him and a look of great wonder slowly spreading over his
pudding face while Miss Grey gave him a perfunctory scrub
with a blackboard duster and then dashed off loud in the
pursuit of a box of mutton-pies. With suchlike diversions
we passed a very happy morning.

All pretence of work stopped at noon when we were
set free for an hour's play in the open. It was the one time
in the year when even a game of football could not hold
our interest. Mothers and sisters had begun to arrive with
exciting parcels. Our young bellies yearned for the spread
to come. We made backs for each other—I was always the
base—so that we might look through the windows, but they

were so clouded with steam that we couldn't see anything. Our appetites mounted on denial, till we could almost see the laggard minutes dragging themselves across the leaden sky. It was a bitterly cold day, with a light but sharp-edged wind that searched out the marrow in our bones to kill it. Standing at the top of the playground we could see all the world for miles lying stricken bare and cheerless in the grey chill of death. The stiff bare branches of the trees held themselves rigid against the wind that sometimes rose into a thin shrill of agony. I lay over the wall with my clumsy boots feeling for a hold between the stones, and looked down the hill towards Dungair. The grey farmhouse seemed very desirable as it stood in the lee of the steading with dark smoke streaming from its chimneys. I knew that smoke. Sally had just made up the kitchen fire after dinner by heaping on an armful of peats and covering them with dross from the coalhouse. In an hour the hearth would glow like Daniel's fiery furnace and she would bake the scones for tea. As I watched and longed to be warm beside that fire (perhaps Sally would sing her only song for me) the farmhouse grew dim. A veil came down between us, blotting out the farm as it had never been. Quite breathless I watched the land disappear as if it were crumbling away at the Day of Judgment, nearer and nearer, till the houses of the Stripe in the middle distance were gone too, nearer and nearer till the wind suddenly shrieked most eerily in the trees and chilled me with an ultimate unendurable cold. Then I felt a bluffert of something in my face. The snow had come.

Something had happened at last to break the intolerable discipline of waiting. The school went mad. We left the shelters and porches, we deserted the doors and windows, we raised a communal yell of primitive delight and rushed about the playgrounds madly, pushing over the girls, trying to catch the snowflakes in our hands and even rolling on the ground in our ecstasies. The air filled with shrieks and yells and driven snow and the inhuman whistling of the wind. Suddenly the bell rang. The din halted for a moment of intense silence, for even the wind seemed to hold its breath. Then we all rushed for the schoolroom and the Feed.

The next four hours were the one time of the year when the school wore any aspect of true humanity. The sliding partition had been opened, throwing the two rooms into one, and revealing the total splendour of the decorations. It was a riot of coloured paper, evergreens and holly berries, and there was no contretemps, as on the occasion when one young devil adorned the portrait of Queen Victoria with turnip tops. The big table in the Infant Room bore a noble array of tea-urns and cups, while another table supported a rare assortment of fancy cooking. Odd hampers stood about, full of luscious, exciting things that would be revealed in due course. It was with the greatest difficulty that we could be persuaded into our places. When we had been reduced to order, our mothers and sisters found seats on the benches where most of them had learned the A B C in their day; Miss Thom opened the piano; Miss Grey banged the floor with her pointer; and the company rose and burst into full-throated melody about 'When Humble Shepherds Watched their Flocks'. After the angels had been assumed back into Heaven, with a distressing uncertainty as to tone among the older boys, we chanted the Lord's Prayer with unusual heartiness and then broke into a clatter of excitement. Two ladies appeared with the tea. Others distributed bags of cakes. The Grand Feed began. Each child received a bag of buns and a mutton-pie which had to be eaten before the more interesting sweets came round. We were not fastidious. Our rural appetites had not been corrupted by éclairs. We fell on the rather sober pastries with happy shouts and washed them down with cupfuls of hot, sweet tea. It was a heartening sight, with sixty feeding as one, like a byreful of nowt. When we had finished the plain fare and burst all the bags, after much misdirected expenditure of wind by the little girls, the mothers and sisters handed round their baskets of home bakeries—jam puffs, almond cakes, shortbread and sugared biscuits. And still we ate and still the wonder grew that we didn't burst. Nature beat us at last. We had to give up. Those who could had now to wait on the older people while the others retired to the lavatory and were sick.

After the tea was over, a few of those Christian old women who apologize for their presence at public festivals by doing all the work, cleared away the dishes while the rest of us—scholars and guests alike—embarked on a programme of amusements, dictated by tradition and rigidly enforced by Miss Grey. (Perhaps I may seem a little hard on Miss Grey, but she is such a noble spirit that she demands to be presented, warts and all, like Cromwell, whom she resembles in many ways.) I forget the exact order, but I will never forget the amusements, for I enjoyed them at eight consecutive treats. We sang 'Willie's gane tae Melville Castle', 'The Auld Hoose', 'The Lass o' Gowrie', 'Good King Wenceslas' and 'Scots Wha Hae', We played the 'Farmer in the Dell' and 'Forfeits' (yes, with kissing, at which the older boys breenged like young colts); 'Nuts and May' and 'Blindman's Buff'. We danced eightsome reels while Miss Thom put all she knew into the left hand, and some tried what they thought was a waltz. And we said our pieces, very shy and very conscious of our female relations, sitting bunched up on the quite inadequate seats. I said a piece which began 'What does little birdie say?' I knew it was a silly piece because although I was (and still am) unashamedly ignorant about the habits of birds, I did know that they never said anything at all intelligible at daybreak or any other time. All the same, I said the piece because I was always willing to please. Indeed, I must have been a pleasing sight, a small, incredibly fat child, bursting from my last year's suit and well plastered with jam-puffs, wheezing out the vapid poem with my back to the wall in every sense of the phrase. An imaginative spectator must have concluded that the centuries had not laboured in vain when they had produced such a ridiculous but well-covered *mus*.

The afternoon flew by in well-disciplined revelry, and the snow piled itself deeper on the window-ledges. The school grew dark. Miss Grey lit the paraffin lamps and heaped up the fires with an expense of coal that would have stricken the hearts of the parsimonious school board. The atmosphere was dim, warm and smoky. There were sweets and fruits and crackers for all. Time passed unheeded; on this one afternoon no weary eyes watched the hands of the clock. We would have been content that school go on for ever. Indeed, it might have, for the storm was now so furious and the snow so deeply piled on the roads that we could not have gone home along the dark woodland paths. Perhaps we would have to stay in the school, lapt in jam-puffs and crackers, till the storm had spent itself. The day began to wear the aspect of a great adventure, But we were disappointed. Parents began to arrive at five o'clock with gigs to take us home, and so we had to go, unwillingly from school, but we minded the less because we had filled our pockets with the fragments of the feast. Dungair came for me in the gig. He was an enormous white snow-man, with drifts in the lurks of his collar, the pony was whiter than a winter's ghost, and the gig like a coach in a Christmas pantomime. The storm blew over as we drove home. The moon came out and the world lay still in sleep under the gentle snow. And before we got home I was asleep too, dreaming of a disordered heaven composed of cherrycake and marzipan.

That was a remarkable winter, because we had not only the treat but also the magic lantern. It was some time after the treat that bills appeared advertising a lantern lecture in the school on emigration to the Golden West. I can assure you it was an event in the naïve world of 1914 when Hollywood was yet undreamt of. We all attended one dark night, for although the date had been fixed for full moon nothing could be done about the clouds. We packed the schoolroom—men, women and children—packed it in a way that was no doubt very exciting for the adolescents but very uncomfortable for small boys jammed amongst mountainous elders. When the school could hold no more, the lights were put out, the lantern began to hiss and the lecturer appeared. He spoke with great fluency about the Golden West, but I do not remember anything that he said, nor even what he looked like. I do remember the way he banged his pointer on the floor as a signal that the operator should change the slide, and the roars of laughter with which we greeted every one that he put on upside down. I suppose it was rather a dull lecture and that we were glad of any comic relief whatsoever, for we broke into riotous laughter when he burned himself with the lantern and swore as from the heart. The Golden West should have appealed to me, for emigration was the only way in which I could have hoped to better myself when I grew up. If I stayed at home it was unlikely that I could be other than a farm servant. Education beyond the Board School stage would have been out of the question, for, in strict justice to their own children, my grandparents could not spend their little money on me. Therefore my noblest prospect was the road to the Dominions. However, the Golden West failed to excite me, which was perhaps as well, for I have not yet had to emigrate to it, for either my own or my country's good.

As the first days of spring came in the Old Man, now in his seventy-seventh year, felt in his bones the nascent glories of that wonderful season. Life was never dull for him, but that spring brought him two adventures of which he was the hero.

The first adventure was The Night on which the House Took Fire. At the opposite side of the kitchen from the hearth there was a recess beneath the staircase in which we kept the peats and into which the maids used to push the sweepings when their mistress wasn't looking. Well, on this memorable night those sweepings had included a live coal which kindled into flame in the many draughts which played about the old house. So my grandmother awakened about three in the morning to find the house filled with smoke. She ran out into the lobby, where, gleaming evilly through the smoke, flames were already eating into the wood of the staircase. She raised the alarm. Sally, who was sleeping upstairs, saved herself by sliding down the ivy in her nightgown. Uncle Sandy appeared from the dickey in his shirt and trousers, roused the men, and organized them into a fire brigade. The smoke, the flames, the cries and confusion were tremendous. Everybody ran about knocking things over. Sally kilted her nightgown and began carrying basins of water to the fire. My grandmother filled her lap with her precious bottles from the lobby press and the men formed a bucket chain from the scullery tap. We had a good water supply so the hiss of steam was soon added to the pandemonium which now resembled a rainy day in Hell.

Our first efforts, which included the opening of every possible door, seemed to augment the flames, and it looked as if nothing could save the house, for the fire was devouring its way upstairs. And just then Sally remembered that the Old Man was sleeping there. She had considerable presence of mind. Instead of raising the alarm, she whispered the awful news to Sandy and quietly took his place in the bucket line while he went off to rescue his father. It took him only a minute to find a ladder and climb up to the window of the room in which the Old Man was sleeping. He burst it open, entered the room—and found it empty. That gave him a nasty set-back for a moment, then he breenged through to the landing. At first he could see nothing for the smoke which stung his eyes. Then his vision cleared, and, outlined against the lowe of the burning stair, he saw the Old Man in his nightgown, standing on the edge of the furnace and coolly pouring into it the contents of the spare bedroom ewer.

Sandy took the Old Man by the arm and shouted—

'Come on, ye aul' deevil, or ye'll be burned tae death.'

'Ach,' said the Old Man, 'I'll burn seener or later so foo nae noo?' with which he flung the ewer into the heart of the flames and danced as a shower of sparks rose up beneath his nightgown.

Sandy made another attempt to draw him off, but he only shouted:

'Damn ye. Can a man no get leave tae enjoy his ain fire,' and then took a firm grip of the landing rail like an old captain determined to go down with his ship.

Sandy wasted no more time, but came down the ladder again and redoubled the work of the bucket line. Inside half an hour they reduced the flames to a mass of steaming cinders. The house was saved. When the Old Man saw there was no more danger he shouted down to my grandmother:

'Woman, gie thae lads a big dram oot o' the bottle i' the clock; an' Sally you lat doon your nichtgoon or the fire'll maybe start again.'

Then with a hearty and accurate spit at a smoking beam he hitched up his nightshirt and retired to bed.

The second adventure concerned Peggy our servant maid. Peg was far above the ordinary run of maids, for, although she was a bastard and had been brought up by a slatternly old cottar woman, she had an innate self-respect which expressed itself in general tidiness and good taste. She was now twenty-one, with straight yellow hair, honey-coloured eyes and a wide mouth which smiled most kindly. Many men had made a pass at her without success.

'Aye, aye,' said the Old Man, 'the seener ye're mairried an' raise bairns the better.'

Peg thought the same, though I am sure she was too proud to say anything about it. She had a lad called Jimmie who had been walking her out for eighteen months but who showed no signs of producing a ring. In other respects he was her ideal: his wooing lacked nothing but specific intentions. Now the Old Man was very fond of Peggy because she was young and good to look on. He knew she wanted Jimmie and he made up his mind she would get him. He cornered the young man once or twice and told him that if he had any sense he would marry the girl. Jimmie hodged from one foot to the other, made confused noises in his throat, but refused to come to the point. Some months later the Old Man forbade him Dungair and so policed the little maid that Jimmie could hardly get a word with her. That made him want her the more; so much, in fact, that he redoubled his efforts to see her, but she, being in the tacit conspiracy with the Old Man, did nothing to help him. Jimmie's feelings got to such a pitch that he determined to take her by storm. One Saturday night he stole into the court at Dungair, propped up his bicycle ayont the kitchen door, took off his boots and climbed in his stocking soles by way of the water-barrel on to the roof and up to Peggy's window. Just short of the window he dislodged a loose slate which slid down the roof and smashed on the cobbles below. The noise roused the Old Man, who knew by instinct what was happening. He rose, slipped out through the gable window into the court where he found the bicycle and the boots, impounded them, returned to his room with them and went to bed. Jimmie, who had been hanging on to the roof in the fear of death, for he knew the Old Man had a shot-gun and might use it, now gave up all thought of Peggy, returned to earth and made off for home, bootless, as fast as he could go. The distance was a good five miles and the roads were hard, so he could have been in no very happy state by the time he arrived there. Next morning the Old Man said nothing about the night's adventure. He just waited. He did not wait in vain. Jimmie waylaid him at the back of the corn-yard a few nights later, and, with a deal of confusion, asked for the return of the boots and the bicycle.

'Na, na,' said the Old Man. 'Onything lyin' aboot the court's mine ' I'll keep it—but, gin ye wis tae mairry Peggy I micht gie ye them for a mairrage present.'

The young man replied that he wanted his boots more than he wanted a wife and that a farmer's son like himself would be shamed by marriage with a misbegotten servant girl.

'Maybe,' said the Old Man, 'an' ye'll be waur shamed when I tell the story o' the boots an' the bicycle in the Corn Market on Friday.'

That gave the young man something to think about. He knew the roar of ill-natured laughter that would go up in the countryside if the Old Man told the story, and Dungair, seeing that he had found the tender spot, probed as only he could probe. When he had got Jimmie into what is known as a state, he said:

'Tha lassie's doon among the hens in the Teuchat Park, an' I dinna think ye'll hae tae speir her twice.'

He didn't. They came up an hour later, the Old Man gave them his blessing and Jimmie went home booted, bicycled and promised. They got married at the Whitsunday term and went to one of the Dominions, where they prospered exceedingly.

Whenever the Old Man told how he had made an honest man of Jimmie, he would add:

'Aye, I doot I'm growin aul'.'

One marriage leads to another. Sally, who had always a constellation of young men about her, suddenly chose the best of them and decided to marry him. When I say he was the best, I mean that he was good looking, intelligent and kind, three qualities in which he differed from all the others. He was also poor, which the others were not; but our family have never bothered about money, and I am sure that the state of his pocket-book was the last thing the Old Man thought of when Peter asked him for his blessing. Peter was the youngest of a large family, an engineer to trade, and spent all his money in seeing the world. I liked him because, though he seldom spoke to me, he treated me as his equal when he did. I thought it very foolish of Sally to get married, but the fact that she was marrying Peter made it seem less foolish than it might have been. And there were compensations. The other louts who had come after her now deserted Dungair. That was a major blessing, for they plagued my life with their clumsy jokes and their excessive masculinity.

Then I too awoke to love. There must be some truth in the saying that the first love is the best love, for this affair at the age of eight was the only one in which I fought for my woman to the effusion of blood. It was also my first, but by no means my last, experience of the eternal triangle. We were all three in the same class at school—Danny, the son of a nearby cottar; Mary, the daughter of a crofter; and myself. Danny and I were friends. He was the first friend of my own age that I had ever had and we had walked to school together in perfect amity for two years. Then Mary appeared on the scene and destroyed our beautiful friendship. Her father took a croft which lay between our homes and the school and she came to school a few days after the May term. We both fell madly in love with her, perhaps because of her flaming red hair. We waited for her at her road-end every morning and took her home again at night, both of us constantly trying to impress her with our cleverness. I had no chance at all during the first week because Danny was incomparably my master at running, leaping, throwing stones and such rustic arts. By Friday he had captivated her under my very eyes. Now it so happened that our places in class were determined every Friday on the results of the week's work. When Miss Thom rearranged us after the dinner-hour, I found myself top, as usual, with—O, joy—the red-headed enchantress second, and Danny, as always, at the foot of the class. Thus I had the right to sit beside Mary until the next Friday. Having gained this advantage I consolidated it, and by prompting when her own wits failed her, was able to keep her in the second place for the next two weeks. So the game went on—Danny fighting with his lovely muscles and I with my dishonest wits to win the lady. Then came an afternoon when Danny was kept in for some misdemeanour. Mary and I went home alone through the sunny woods, played in the burn and hunted arnotts in the loamy brakes. We were still playing at her road-end when Danny appeared after his penance in the musty schoolroom. Three-quarters of an hour of arithmetic had not improved his temper and my success with Mary had not lowered my pride. He tried to pick a quarrel and I, forgetting my dislike of trouble, gave him back word for word. Words led to blows. He punched me in the belly. I hit him in the eye. We wrestled, bit, kicked and scratched, while Mary sat on the dyke and squealed with delighted horror. Then I hit Danny on the nose with my knee. A stream of bright red blood broke over his mouth and fell upon his boots. I stood back gasping; Danny clutched his nose with his hands, then held them out all bloody; Mary gave a terrified screech and flew off home, leaving me alone and master of the field. It was an empty victory. She did not speak to either of us till after the holidays when both of us had forgotten her, and we never admitted her to our friendship, which became more beautiful than ever.

So the spring came sweetly in, with lengthening days and softer winds. The rising sap made the young people think

of marriage, but it also stirred the Old Man's rheumatics
and brought him intimations of mortality. By the first of
March the working day in the fields was six to six. Another
sowing was in preparation. With the Old Man finding his
chair more grateful than the brae-set fields, with Sandy
taking over the direction of the farm from his stiffening
hands, and I becoming always more conscious of my world,
the old generation was passing calmly to its night, the new
was coming to its noon and the third was rising slowly in
the east. It had always been so, nor would it ever change.
The Old Man had been each of three generations in turn.
Now he had only to get Sandy married, to bid good morning
to a grandson that would be Dungair in time, and then to
bid the world good night forever. That was the tradition in
which he had lived his seventy-seven years. He was part
of a continuing purpose that was Dungair. That continuity
assured, he could die in peace. As spring grew into summer
he could have had little doubt that that tradition would be
maintained. May was warm and full of the growing seed.
Oats in the Teuchat and Brae Parks, barley in the Moss,
potatoes in the little Corner and hay in the Home Park came
up with the promise of a splendid harvest. In June we hoed
the turnips; in July cut the hay and cast the peats. Rain fell
at the right time and, though there were anxious nights,
the morning never belied the evening's promise. As the year
grew richer and the sun rode high in his meridian, we felt in
our bones that we were living through a year of memorable
beauty in which tradition would attain its finest flower.

But in June there arrived at Dungair a portent of evil
omen—Cousin Jonathan's new Ford car. Cousin Jonathan
was the son of my great-grandmother's youngest brother.
He had left his country in the middle 'eighties, had adven-
tured in Australia, South Africa and Alaska—wherever
gold or other precious metals offered a quick reward for
easy labour. He had made a few fortunes and spent them

quickly, but he was one of those whose spirits have been so quickened by the touch of gold that they never lack for money. In 1914 he was a fairly prosperous fruit-grower in California, had married a sweet Austrian wife, and came back to Scotland on holiday, bringing his Austrian lady and the first Ford car ever seen in our part of the country.

Jonathan and the Ford would have made that summer memorable in themselves. He was a fine figure of a man, big, red-faced and jolly, always telling stories of the deserts and the Yukon Trail. The Old Man took a great fancy to him and honoured him with the rum bottle from the bottom of the clock. I am not sure that the Old Man did not like the Ford car even more. He certainly developed a passion for motoring and Jonathan could not drive fast enough to please him. Jonathan was a reckless driver at any time, but, inspired by Scotch whisky and urged on by Dungair, he rode the Ford along our narrow ways like a juggernaut. Full of whisky and the lust for speed, these two elderly gentlemen tore around the country in clouds of dust and fumes, killing hens and ducks, alarming the cattle and scaring lazy cart horses out of their wits. My grandmother remonstrated with them. They would kill themselves, she said.

'Damn't,' said the Old Man, 'if I've got tae dee some day I micht as weel dee in style.'

My grandmother then tried to get the charming Austrian to use her influence, but Fritzi just shrugged her shoulders and made a hopeless gesture which, I have no doubt, my grandmother perfectly understood. Both had married men with minds of their own. So Jonathan and the Old Man terrorized the countryside at their own sweet will all through the month of June. But the thunders of that stinking car were the death-roll of their pleasant world.

July 1914. Our social year came to a splendid climax in that enchanting month with the School Picnic, the Salmon Tea and the Games.

When I returned to Dungair in search of my own tradition I found an album full of photographs of our family on
their social occasions. Among the constrained groups, posed
against incredible backgrounds and oddly contrasted with
them, were a number of prints of the School Picnic—lumpy
boys and frilly girls in gaily decorated farm-carts, drawn by
horses smothered in ribbons; parties of young wives nursing
their babies on the sands; and one—O, noble sutlery—of
Dungair himself presiding over the whisky jar. Some of
them went back forty years or more; and I could trace half
a century of social life in them, for the Picnic was a major
event in our life before the war.

It was always held on the sea-shore in the first week of
July. Although it was intended for the children, the whole
family—parents, dogs and all—used to attend. It was a
communal holiday to which every one contributed something. Dungair, for instance, would send two carts and a
stone of apples; Uncle Scott sent one cart and a sack of
sweets; the ladies provided the tea; the twins from the
Mill supplied the music with their fiddles; and everyone
brought their own mug. Preparations occupied a full week
at Dungair. The carts, covered with a winter's mud, had to
be washed and painted. Harness was cleaned, chains and
buckles polished to silver, and all decorated with ribbons,
coloured straw and pom-poms. My grandmother baked
sausage rolls and sweet cakes for days, little round cakes
in goffered tins, squares of shortbread, and heavy slabs of
cherries and marzipan. The prospect of so much sweet food
raised me to a high pitch of excitement by Friday night, and
Saturday never failed my expectations.

On the first Saturday in July 1914 a gay cavalcade set out
from Dungair under the bright morning sun. Jock, the foreman, Sally and myself rode in the first cart; Tom, the second
horseman, in the next; and Sandy in the third. The carts
shone red and blue after the painting. They were festooned

On the way to the Picnic

with gaudy streamers, bannerets flapped bravely at the corners and the first cart—my cart—flew a Lion Rampant, red and yellow, two feet square, from a stout round pole. The horses shone too with their gleaming coats, their glinting chains and the wealth of decoration on their harness. It was, I thought, a royal show; and Jock must have thought so too, for Don, the bay gelding that drew our cart, bore a magnificent crown of blue and white rope and satin above his saddle. We shone, our equipage shone, the whole world shone that summer morning as we set out for the school where we should pick up the rest of the company.

We left the school at noon, twenty farm carts loaded with delirious children and their elders still a little self-conscious in their almost Sunday clothes. The Old Man led the procession, driving with Uncle Scott behind the white pony. They were two magnificent old gentlemen in their square white beards, square bowler hats and ruddy age. The first Dungair cart followed with its satin crown and Lion Rampant and the rest fell into line all anyhow. The commissariat, presided over by some of the ladies, brought up the rear—last but by no means unthought of. The whisky jar travelled in the gig with Uncle Scott and old Dungair, of course. Miss Grey, like the good general she was, had gone ahead with Jonathan in the Ford to make her final dispositions on the field.

We enlivened the morning with cheers and song. If that elusive instrument, the welkin, ever rang, we rang it that forenoon. Every farm we passed drew a loud salute from us and even a stray hen on the road received her three times three. The older people soon lost their self-consciousness and bawled with the best of us. We let the world know we were on holiday. Miss Thom, the infant mistress, was always charming at those picnics: as she taught the babies to read and write in school so she now taught them how to enjoy communal pleasures. I can still see her as she rode in the Stripe long-cart, under a canopy of streamers and

coloured lanterns, surrounded by twenty children from
the infant class. She sat on a box in the middle with the
little dears all about her in various states of excitement
and misery, like a symbolical tableau of Motherhood or
Britannia. And indeed Miss Thom was a mother to them
that day. She upended them when they fell, eased them in
their uncomfortable dresses; told them when to cheer and
how, and, most important of all, wiped their noses which
is the first condition of pleasure at five years old. We owed
a great deal to Miss Thom for her kindness and patience
when we were very small, just as we owed perhaps more to
the invigorating discipline of Miss Grey when we grew older.
Miss Thom sweetened the transition from home to school;
Miss Grey taught us the respect for the Lord which is the
beginning of wisdom; together they made as much as any
one could have made of the rather poor material at hand.

We arrived at the seaside about one o'clock. Miss Grey,
Jonathan and some of the ladies were waiting for us,
superintended the pitching of the camp and sent us about
the business of enjoying ourselves. We always made a rush
for the sea; off with our shoes and stockings and plunged
bravely into the first gentle-looking wave. Some of the
older boys even withdrew along the shore, stripped and
dooked in a state of unrighteous nudity. I was never one of
them, for I never liked the sea and never will. I contented
myself with dabbling my toes in the little waves until the
bell called us to eat. Then, and only then, the picnic began
for me. There is no need to tell you how we ate and drank,
played games, sang, danced, ran races and paddled in the
sea, while the salt wind blew sweetly over the thyme and
the sun sped across the cloudless heaven. As the whisky
jar yielded up its inspiration the fathers grew jolly; their
mood communicated itself to the women; Dungair would
challenge them to wade; and after a great deal of laughing
they would accept the challenge. That was twenty years ago

when modesty was rated higher than hygiene and women were not supposed to show a leg in public. So you can imagine the excitement with which the douce old wives kilted their coats, took off their stockings, and ran self-consciously down to the sea like fantastic birds. And you can also imagine, I hope, the comments with which men like Dungair greeted their stout white legs, legs unnaturally white, like stems that had been too long in the dark, and the coarse innuendoes which shocked and pleased the ladies.

It must have been a diverting sight to see our pagan naturalness breaking through the overlay of convention, but I was too young to appreciate it. What I do remember is the noble prospect of Dungair and Uncle Scott, those square white-bearded gentlemen, with their boots hanging round their necks by the laces, their socks stuffed into their pockets and their trousers rolled up above their knees, standing a little into the water and regarding the ocean with an expression of contempt, like a twin Canute, confident of mastery.

Towards six o'clock Miss Grey rang the bell, we finished the remains of the food and prepared for home. Miss Thom collected her infants, gave their noses a final wipe, set their clothes to rights and settled them in the long cart. We all got aboard after a deal of confusion and, led by Miss Grey and Jonathan in the Ford, we raised a great cheer and set off for home. We were still a jolly company, but tired with playing and heavy with the sea air. Our songs and hosannas were lazy with contentment; they drowsed away into a fine echo of their morning glory; and at last they were only a thin music fading over the countryside where the dew was settling as the sun went down behind the hills and the air grew cool. So we went home to Dungair, I between Sally and her Peter in the bottom of the cart with my sleepy head in her lap. The harness jangled, the cart rocked on the uneven road, the Lion Rampant fluttered in the light wind and Sally sang her only song:

There's many a bonny lass
In the Howe o' Auchterless
There's mony a bonny lass in the Garioch-o
There's many a bonny Jean
In the Toon o' Aiberdeen
But the flooer o' them a is in Fyvie-o.

All the happiness of my little years is entwined with that old song.

In contrast to the Picnic, the Salmon Tea was an intimate function, but none the less delightful for all that. Our host was Mr. Swan, a lovely old gentleman who owned a large farm at the seaside and was also lessee of the adjacent fishing. Being a hospitable bachelor he used to invite a large company of his friends to tea on an afternoon in July, a tea at which the *chef-d'oeuvre* was a twenty-pound salmon. I enjoyed the teas, not only because of the salmon, but because Mr. Swan's farm had a number of exciting features which I was allowed to explore at my own will. For instance, his mill-dam had a little artificial island in the middle on which the saughs grew down to the water and made a green covert where a small boy could lose himself for a whole afternoon. Then there was a peacock in the garden whose tail filled me with wonder and awe, not only by its beauty but also by its disproportion to the size of the bird. And there was an ice-house, a gloomy cavern deep in a hillside where the ice used to be stored before the days of modern refrigeration. That ice-house was perhaps the most exciting thing in my young life, for Mr. Swan, who had a sense of humour, used to see that a few bones were lying about in it when I was likely to pay a visit. Not only did I sup on salmon that day but I surfeited on horror. Bless the shade of Mr. Swan wherever it is. The good old man is dead these many years, but I will never forget his twinkling eyes, his ruddy face and his Vandyk beard. He gave me sixpence every time he met me and thus laid the foundations of a Savings Bank

Account which I tended with all the care of a French peasant till I was twenty-one, when I spent it most satisfactorily in one night of glorious living.

The games fell on the first Saturday in August this year. Their advent quickened the whole *tempo* of our life at Dungair for a fortnight, for the men indulged in what might be called an orgy of athletics. You see our games were an intimate affair. As the prize money was small, the big men like Dinnie and Cameron did not compete; the field was left to men of the district, and many local feuds over the wenches were continued at the heavy hammer and the wrestling. The Old Man had been an athlete in his day; he had been incomparable at tossing the caber (at our games) for many years; and he liked that some of our men should enter for the games in order to maintain the reputation of the farm. So when July came round he used to look out the fifty-six pound weight from the barn, take down the caber from the loft, and challenge the men to do their best with them. They practised every evening on a bit of grass outside the bothy door—the Old Man sitting on a box and making derisive remarks about their lack of virility while they hurled the weight a distance quite out of proportion to the energy they expended. However, they learned through practice and usually one at least was good enough for the games. On this occasion Tom, the second horseman, though he never tried heavy athletics before, showed a certain aptitude with the hammer and the Old Man began to treat him with something less than the usual contempt as the day of the games drew near. The rest of us were sure that he would win the first prize and I desired that honour so fervently that I offered prayer three times for his success.

The day of the games dawned modestly but cleared up to a brilliant noon, with the heat of the unclouded sun tempered by a gentle wind from the sea. The old people and I drove over to the games field in the gig and as we came through the wood we could see the tents on the level haugh beside the river, the bright colours of the dresses and the fantastic tower of a merry-go-round high over the heads of the crowd that were already gathering about an oval in the centre of the field. We unyoked the pony under a vast beech tree, hitched her to a post and went over to join our friends. I wish I could set down on paper, as I can see so clearly in my mind, the rural company who met that day, so secure in the enjoyment of their pleasant world and who were never to meet again in that or any other place this side of death. I wish I could put on paper and make, as it were, eternal, that scene which then seemed in the nature of abiding things; but too much water has run under the bridge and time has rubbed a bright veneer over the remembrance of things past so that, try as I may, I can recall no more than painted ghosts from the dead. I passed the games field the other day. The sun was shining as it shone twenty years ago. Dry moss was still growing in the seams of the bridge, the river was still breaking in silver over the stones and the trees still stood round the field in ample majesty. So much was unchanged—but so much more was changed out of all recognizing. It was not only that the people were dead or grown old, but also that I myself had read too much and seen too much since then, so that I had lost the fine edge of memory. All that I had left was a blurred imprint of beauty, for the years which had taken so much loveliness away had taken the ill as well. But, if I can set down even a little of that remembered beauty on paper I will feel that I have been able to save some small part of my world.

This is the old world as I remember it—farmers like Dungair and Thomas, slow-footed and bent under a weight of years; my grandmother and Aunt Scott in black bonnets and black-trimmed capes, like pyramids in mourning, with wisps of white hair breaking from behind their velvet bows; young men like Uncle Sandy, in high waistcoats and higher collars, rather dashing in a clumsy way and smelling of hard yellow soap; handsome girls like Sally in yellow blouses and long skirts that swept the grass with incomparable and unconscious grace as they ran about in threes and fours; farm boys in their very tight Sunday trousers which made their legs seem quite inadequate for their slouching shoulders and their clumping feet; servant girls snatching a nervous pleasure by flirting with the men before the faces of their employers; and children everywhere, from babies to hulking brutes in knickerbockers. Some leaned against the ropes, some sat on forms, some lay in the grass, some walked slowly up and down, constantly stopping to greet a friend or pass a joke. A murmur of gossip and laughter, mingled with the shouts of the children, rose on every hand, while the sellers of candy and apples sang a song of 'three a penny' and the merry-go-round turned a sweet sentimental tune into acid drops of melody.

Three large tents stood in a shady corner of the field. One was for the Committee, one for the competitors, and the third for refreshments. As the Old Man was on the Committee we went to the first at once and the Old Man began to revive the glories of the days when he could toss the caber further than any other man in three parishes. Such memories were dry work. They demanded the refreshment of many drams, which led to a contest in another rural art, the drawing of the longbow. The old men would have argued all afternoon, but time was passing and, after the secretary had run himself almost off his feet, he managed to get the Committee out of the tent and the first competitors on to the field.

The next three hours saw a desperate but intimate struggle among the young men of the parish. They ran, leaped, threw hammers, tossed cabers, wrestled and cycled against each other, while their friends cheered them or shouted dry but not unkindly criticisms when they failed. To our great delight Tom threw the hammer a prodigious distance; 'Donal' Dinnie couldna a deen better in his sleep, man'; and won the first prize with yards to spare. He and the other prize-winners became the heroes of the afternoon and, as soon as they drew their prize money from the Secretary, they made straight for the refreshment tent and the stalls, where they spent it on drinks all round and fairings for the girls. They owned the world that afternoon, and as evening came on the swagger of their shoulders told old men like Dungair that a new generation had taken their place and that there was nothing left for the old but a seat at the fire and then good-bye.

The games finished between five and six and those who had bicycles went home to supper. Those who had not, or who could not bear to leave the field, ate sandwiches in little groups under the trees or bread and cheese at the refreshment bar, and all washed down their meal with bottled beer. Many of the young fellows were already a little drunk and their careless spirits kindled a recklessness in the girls. Little squeals rose from the parties under the trees and some of the wenches even made so bold as to drink a few mouthfuls of beer out of the bottles, to the accompaniment of choking and laughter. Meantime the Old Man and his friends had some settling-up to do in the Committee's tent while my grandmother and the other ladies were accommodated with port wine and shortbread in the now empty competitors' marquee. When the Secretary had made sure that every cork had been well and truly drawn and when some of the Committee were beginning to stagger under the load of their responsibilities, the Old Man collected my grandmother and myself and we went off to tea with a friend nearby. As we left the field in the gig, the young people were returning from supper, ready for the evening's fun. A dancing board had been laid under the beeches and a small band consisting of a fiddle, two cornets, a clarionet and drums, was tuning up for an eightsome reel. Meanwhile the sun was casting long shadows over the field and its light was tinged with a faint gold in transit to the rose, as afternoon passed into evening and the fiddle's thin music grew upon the air.

We visited the field again on our way home, for the Old Man could not pass the place where he had drunk so often and so deeply. The evening was very still and the music, augmented by the incongruous sounds of the merry-go-round and the drummling of the river, came loudly across the field while the feet of the dancers beat a heavy insistent rhythm on the board. That was a night to remember the old world by. The sun had gone down. Shadows were drawing in from the wood. Flares hung before the booths and lit a constellation of stars among the bottles in the refreshment tent. The young people and some who were not so young had gathered in well-defined blacks of shadow about the rings of light at the merry-go-round, the dancing board and the stalls, while some maintained a traffic to and fro between the entertainments, or stole away arm in arm into the privy wood. Others who had drunk wisely swaggered about with wide sweeping motions, now trying a snatch of song, now roaring out a challenge to the world, masters of their fate for this one night; while a few who had drunk too well lay fast asleep where they had fallen, It was a pagan festival, dedicated to the old traditions of the countryside, and it centred around the dancing board, for the oldest game of all was being played out there. The technique may have varied, but the spirit was the same, Whether it was a tall young farmer with his black hair falling over his forehead as he whirled his partner masterfully round while his eyes burned above her upturned face, or a ploughman butting his wench in the clumsy motions of the waltz, like the movements of a rude stone idol, they had only one thought,—that they were young and the night was short, while the girls could only remember as desperately as they could that life is long and honour fleeting. So it had always been: so it would always be.

We did not leave the gig but drove to the whisky tent where the Old Man took a dram, then we made a round of the side-shows and finally drew up beside the board. I was all exclamations of wonder as I noticed dancers I knew and my grandmother talked to someone who was standing by. The Old Man did not speak at all. He sat very square in the gig, listened to the music of the country band and looked at the board where the couples circled round and round. I wonder what he saw there. Perhaps the shapes of the dancers dissolved and re-formed as they might have been when he was twenty. Perhaps he felt the bitterness of age when youth is at its lusty pleasures. Or he may have thought that they would sometime be as old as he and that they were wise to enjoy themselves while they were able.

Now the shadows were growing darker. It was time for us to leave. The Old Man picked up the reins, called to the pony and we bumped away over the grass towards the road.

As we turned the corner of the wood I saw the games for the last time—the black figures against the lights, the gaudy lanterns on the merry-go-round and the dancing on the board. Then a tree occluded them and they were gone. The pony trotted quietly along the mossy road. The sound of the games fined away into a remote music that died into the muted undertones of the river.

But when I went back there in August five years later there was no music. The War had intervened.

VII
War and Peace

On the Wednesday after the Games, Britain was at war with Germany. The dreams of a few homicidal lunatics in Berlin, Vienna and Paris had come true and murder had become divine in the sacred cause *Pro Patria*. If you called at Dungair on that evening after the games you would not have thought that we were committed to hate and destroy just such quiet country people as ourselves in a million farms along the European Plain. The Old Man would have been sitting in the arbour with Captain Blades and the bottle of rum, my Grandmother would have been sitting beneath the apple tree half asleep over the *People's Friend*; Uncle Sandy would have been lying in the lythe of a dyke somewhere out on the farm; Sally and her young man would have been down in the wood; and I would have been in the bothy with the men. Not only 'I would': I was. If I had known how important that day was proving for all of us, every event of it would have been burned into my mind, but this alone I remember, that I lay in the grass outside the bothy with the sunset warming my bare legs and read a Wild West story while the men tried to equal one of the hero's feats by vaulting over a six bar fence. They had rigged up two poles with a transverse bar at the five foot measure and were trying to jump it in their stocking soles. Only Tom, the second horseman who had been so successful at the games, was able to clear it and I remember the set expression of his bony face as he poised himself for the quick run and leap. I also remember the laughter when the foreman, muscle bound and slow but willing, knocked down the bar with his lumbering bottom that he could never raise far enough from its native element the ground. And I will never forget how the heat haze grew blue and blue down and along the valley until it merged into the hills that stood against the west like peace become eternal.

Next day Bill and Tom, being members of the Territorials, were called up and went to report at their local depot. I did not understand what it was all about, but I did know that something untoward had happened, for Sandy had gone to town and, as there was no work doing, it might have been a Sunday. The afternoon was dull, thunder hung heavy in the air and the sea boomed as it might have been the undertone of war. I played about the steading half-heartedly, tried to read a novelette in the bothy, did this and that, but found nothing that could hold my mind for long. All the time the red cock kept crowing on the top of the peat stack where he scraped among the dross.

Bill and Tom came home in the evening for the last time, packed their kists, bade us all good-bye and went to the war. The Old Man and I watched them as they cycled down the road. Rounding the corner, they turned and waved to us.

'Puir damned eediots,' said the Old Man. 'That's the last o' them.'

We missed them badly for a week or so, as the harvest was coming on, but we found others to take their place—an old man who was past his best and a halflin loon whose wits were none too good. So another harvest began at Dungair.

It was the last harvest that my grandfather was to enjoy in the old style. Thereafter the broken weather and the shortage of labour made art impossible in the wearing struggle against time. And so I should like to remember that harvest, the last gift of peace to the old world, as the

In the rickyard

Old Man gathered it home. He was long past labour in the fields; he could no longer swing a scythe for long afternoons in the fiery sun, but he had some little arts which he could still practise to his heart's content. At the gable end of the house there were two posts topped by a transverse bar on which he used to sharpen the binder blades. Every evening when the sun was dropping into the arms of the hills and flooding the valley with a soft red light, he sat at the gable and sharpened the blades with a stone. As it was a delicate job he took off his hat to it; the wind played with his thin white hair; the sun touched his beard with rose and the blue smoke from his pipe curled up before his face. He whetted the blades carefully, every now and then testing their edges with his thumb, and so moved down the line, utterly absorbed in his work. When he had finished the job he put away the stone, walked down the whin road to the iron-gate whence he could see the whole farm, and stood there for half an hour in thought. Then he stumped up the whin road again; sometimes he stopped and leaned on his staff with his free hand behind his back, looking over the fields; and so he came in to bed.

It was at the leading of the crop that he showed himself most the artist. In the neuk of the corn-yard where the farm-road left the highway it was our immemorial custom to build the master stack. Every stack of course was more or less a work of art, for the Old Man would never have tolerated those drunken rat-riddled things that some farmers are content with, but the one in the corner, facing the highway and standing as it were the symbol of Dungair, had to be supreme. The Old Man had built many master stacks in his day, stacks that had won him an unchallenged reputation in the countryside, but now he was too old to clamber about among the sheaves and his knees would not bend to that severe discipline. He had to content himself with being the architect and directing the work. So he walked around with a rake in his hand and watched young Sandy build the stack. If a sheaf seemed out of place he would push it in with the rake, every now and then stepping back to see that the walls were plumb. As each cart was emptied he groomed the stack with the rake, drawing it over the sleek wall so that no sheaf might be disturbed. Then he gathered the rakings into a heap to be put away in a safe place where his mortal enemies the ducks might not get at them. The building of the master stack took a long time, because Sandy had to come down for so many consultations, but it was a rite which must not be hurried. As the building passed the eaves and drew it towards the peak, the work grew slower still. The Old Man directed the placing of every sheaf; Uncle Sandy often lost patience and arguments ensued; but the stack grew towards a shapely apex and the last sheaf was tied to its place at ten o'clock. The men went off to the stable, Uncle Sandy went to see to the cows, and the Old Man and I were left alone in the stack-yard. He groomed the rick again, collected the rakings—only a handful this time—then stood back to admire the work from every angle. It satisfied him at last. He laid down the rake, lit his pipe, took a survey of the weather and went in to the house. Another harvest had begun with honour.

We won that harvest in ideal weather. Rain fell sparingly, a dry east wind blew in from the sea to set the ripe grain rustling in the sheaves, and every night the great red moon hung beneficently like a jolly god in the unclouded sky. Sally forked that harvest, working from seven in the morning till the moon was high. I was only a child then, but I have an enchanting memory of that girl as she tossed the sheaves into the carts, while the sea wind blew her skirts against her strong body and the warm blood deepened the dusky rose of her face. It was the last the Old Man was to see of her, for she married at the end of harvest and went abroad with her man; and it was the last harvest he was to enjoy in the old fashion of Dungair; therefore how right it was that the sun and the wind and the grain so blessed him and that his own child moved so beautifully before him, the very breathing incarnation of all that he ever truly loved.

As Peter had got a good job on a South African Railway, he and Sally were married in October and sailed from London three days after; and with the marriage may be said to have come the end of a wonderful year. By every circumstance the wedding was destined to be a memorable feast. In the first place the Third Trust Deed had come to an end. The Old Man was his own master and had actually good money in the bank. In the second place he had always loved feasts and as he was growing old this might be the last he would ever attend. In the third place it was also harvest home and had been one to dream on. And last of all, the war, which had been going on for three months, showed no sign of making an end; a dreadful uncertainty was beginning to trouble our minds; already young men we had known had been killed; and those that remained felt a need to make the best of the days that were left to them. The Old Man made up his mind that Sally's wedding would be a memory to cheer him through eternity.

It took place at Dungair. Preparations went on for days. Miss Betsy, the sewing woman, worked in the dining-room for a whole month making dresses for the wedding, for the honeymoon, and Africa. My grandmother baked for hours on end, until the sweat stood out in little moons all over her nose and dropped into the shortbread, to give that delicacy its exquisite lightness, as she said. Boxes of fruit arrived, and cases of bottles that the Old Man took into his charge. Then the wedding cake in a tea box that had the dry fragrance of the perfumed East. Then the greybeard, a five gallon jar of whisky which the Old Man cached beneath his bed and watched over as if it had been his one ewe lamb. On the morning of The Day everything was ready.

The company assembled in the dining-room at three o'clock. As if indeed to celebrate the end of an epoch, Dungair had invited all his old friends, some of whom were like himself, farmers of the third and fourth generations in those parts. He sat in his easy chair in the corner between the window and the fire with the greybeard between his feet and a tray of glasses and a jug of cold water from the pump on a small table at his side. As the old men came in with their wives, they walked over to shake his hand and he, without rising, gave them a stiff dram from the greybeard while the maid helped their ladies to a glass of wine. They then sat down in a wide circle around him. The younger guests, friends of Sally and Peter, disposed themselves in odd corners and laughed among themselves while Peter stood in the middle of the room and was congratulated after the coarse fashion of the countryside by all who came in. The minister arrived last of all and, after taking a little wine, turned to Dungair and said they might begin.

Sally came in with her mother, wearing a plain tweed skirt and a white blouse on which she had pinned a black cat that Tom the second horseman had given her at the games. The company stood up, the minister took his position on the hearthrug with Sally and Peter before him, the Old Man made sure that the greybeard was safe between his knees, and the service began. We sang the Hundredth Psalm, led by Uncle Scott, whose rolling bass was like the elemental let loose in that crowded room. Then the minister prayed and began to read the ancient and somewhat pagan service. When he asked who gave the woman to be married, the Old Man replied from his chair with a wave of his staff, Sally and Peter swore to love and cherish each other for the good of their souls and the comfort of their bodies, the minister blessed them, and so they were married. It was usual for the company to have to endure a long discourse on the more refined joys and the less obvious responsibilities of marriage, but it was cut mighty short on this occasion, which was just as well, for the Old Man was quite capable of putting a sudden end to the address if it had lasted one minute over the five. So the minister talked briefly if not to the point, the papers were signed, the benediction pronounced, and the healths of the man and wife were drunk in noble bumpers.

The interval of half an hour until the feast was ready was spent in amiable discourse, The younger guests walked in the garden, Dungair and the old men went for a round of the steading, the women looked at Sally's clothes, and my grandmother drew order out of the imminent chaos in the kitchen.

We sat down thirty in the dim gold drawing-room around a board spread with my grandmother's patterned linen and decorated with the last flowers from her beloved garden. The Old Man sat at the top of the table with Mrs. Lambert and Mrs. Brown, two of his oldest friends, on his right and left, and the rest of his neighbours beside him. My grandmother sat at the foot near the door where she could keep in touch with the kitchen. Sally and Peter were on either side of her and the rest of the young people were at her end, for her sympathies were always with the young against the cantankerous interference of the old.

How can I describe that feast? I have no words to set down the ripe humanity around that table, the wrinkled, whiskered faces of the men, the faded primness or the rosy pink exuberance of the women, the dark and exquisite flush of beauty on the young bride's face, the slow Scottish wit; the roars of laughter, the steam of food, the clink and clatter of the dishes, the thin light of evening deepened in the gold of whisky or kindling a dim radiance in the wine—these and the suns of long gone summers that lingered in the faded gilding of the room. I can only say that it was such and such and hope to evoke from your own past the warmth and laughter of humanity.

To me it was very heaven. I had never seen, never imagined there could be such a wealth of lovely food. First, soup with sherry; then a turkey trimmed with bacon and three hens wreathed in sausages; then a giant roast with sauce; then cold meats in abundance; then sweets, dessert and chocolates; and birse tea last of all. Good drink accompanied them—port and sherry wine for women and boys and whisky from the greybeard for the men. So the Old Man feasted his friends at his daughter's wedding and, when the cloth was drawn, hardly a one of us was sober though none of us was drunk.

The dark had fallen. Lamps were lit. The night was growing cold. Sally dressed herself for the road and prepared to take her leave. That was soon over. A few women wept, but they were guests who had drunk too well. The Old Man gave Peter a final dram. My grandmother kissed Sally and said good-bye. They ran out of the hall, out of the house, into a car and drove away.

The Old Man listened till the car ran out of hearing, then he returned to the greybeard.

'Weel, that's that,' he said. 'Lat's mak a nicht o't.'

We did and the fun began immediately. Some of the young men, Sally's rejected sweethearts, had brought fireworks to give her a grand send-off. Very pretty they were too as they rocketed into the frosty sky, but one, imperfectly exploded, fell on a rick of old hay in the Home Field which immediately went up in a blaze. Everybody got tremendously excited except the Old Man. When he saw that the wind was blowing away from the steading and that the fire would not endanger the stack-yard, he made himself comfortable at the drawing-room window where he could enjoy the spectacle with the aid of the greybeard and his friends. This was too much for one of the old gentlemen who protested:

'Dungair, man, ye're fiddlin' whan Rome's burnin'. That's what ye are.'

'Drink up, Stripe, drink up,' the Old Man replied. 'The stuff's weel insured.'

They drank around the drawing-room table until the greybeard gave out and I have one vivid memory of the scene. The Old Men were still at the top of the table, monumentally drunk but sitting steadfast in their chairs, boasting about their youth. The young men, Sally's rejected sweethearts, were consoling themselves at the other end in a litter of empty bottles and orange peel, while one young man who loved her to distraction was sound asleep beneath the table. Tumult and shouting rose on every hand and so the night wore on.

Round midnight the party ended. By good fortune there were enough sober people to yoke the ponies. After many and hilarious farewells the gentlemen climbed aboard, their good ladies followed them and they drove away for home. My grandmother, the Old Man and myself were left alone in the drawing-room where he had a final toddy and my grandmother a little sherry negus. Then we went to bed.

Winter came soon, and on Hogmanay we heard that Tom, the second horseman who had won his hour of glory at the games, had been killed in a night attack in France.

Two months later the Territorials were detailed to defend a senseless forlorn hope. Hardly a man came back. Many fair young men that we had known would never throw the hammer or dance at the games again.

Another year passed over Dungair, a year of growing difficulty compensated by rising prices. If it had not been that so many men were disappearing from the countryside and that labour was becoming a real problem we would have been very well off indeed. We had more money than we had ever seen before and the Old Man became rather unpopular by saying that, though war was damned stupidity for them that fought, it was very profitable for them that stayed at home. As for hating the Germans, that was quite beyond him. Sometimes he got mad with them for upsetting the world, but he made no prayers for their destruction. While the rest of the countryside shouted for blood and took pride in 'giving their sons for Britain', he drew his profit and kept his mouth shut. The war had nothing to do with him. He only desired to enjoy his prosperity in peace; but the prosperity had come too late, there was no peace anywhere except at Dungair, and the Old Man's powers of enjoyment were failing. The harvest of 1915 was a taigled, desperate battle with the weather. Sandy had no time to waste in building master stacks. The Old Man felt himself to be useless and in the way. We all noticed that his step grew slower that autumn and that he stayed closer by the fire.

The Christmas of 1915 saw the end of another old institution—the Sunday-school treat. As we were a good three miles from any church, a Sunday school had been established fifty years before in a cottage near Dungair. This cottage was inhabited by an ancient crone called Susie, a wrinkled, earthy, gnomish woman with a tiny head, a high waist, and many striped and musty petticoats. As she did not require all the cottage, she willingly let one room to the Church. There Mr. Amos Black conducted Sunday school and the minister held a preaching once a month. Mr. Amos was a tall, thin, godly merchant from those parts and had a high Roman nose through which he used to intone God's mercy in a cross between the Anglican whine and the Presbyterian snarl. I disliked his religion, which was too full of prohibitions, but I am sure he was a really good man and sincere in his beliefs. At any rate he was kinder than the God he preached, for he always had a pocketful of pandrops which he doled out to us before he began his exposition of the golden text.

You could have seen us there on any Sunday afternoon from October to June, twenty children between the ages of five and fourteen, all nicely but uncomfortably dressed in our Sunday best. We sat on forms around the low-roofed room while Mr. Black stood in the middle like a father among his children. The proceedings opened with a hymn, one of those maudlin things written for and despised by children. Then Mr. Black clasped his hands where his stomach ought to have been in a man of his age, closed his eyes and, elevating his trumpet nose, told the Lord that we were little children weak in everything but original sin. As the mournful words whined and snarled about our heads, seeking a way to heaven, we sat on the hard forms in all the postures of discomfort and boredom. We could see through the little square window to the hills on the far side of the valley. If it was winter there would be a line of snow along the ridge, or the peaks would be hard-etched against the cold blue of the sky; if it was spring there would be long fires running in the heather; and in autumn we could see the stooks, row on row, like toy soldiers, on the lower slopes; and whatever season it was, the world outside seemed very vast and adventurous compared with our close and uncomfortable walk with God in the Sunday school. We sang another hymn after the prayer; then Mr. Black heard us say the Catechism, repeat the golden text and offer excuses when we had omitted to learn either or both, which I am happy to say was often. Our ingenious lies and excuses pained Mr. Black as much as they pleased us, but he never scolded us very badly and usually ended by giving us an extra pandrop all round. His exposition of the golden text followed, in which he explained the ways of God in terms of the dry goods trade and the ethics of good business. He was a good teacher, for he talked simply and used illustrations that we could understand; but I am afraid that instead of filling us with a desire for the godly life he made us all want to be country grocers. However, I suppose it did us no great harm, and the Sunday school did take us out of the way so that our parents could sleep off their Sunday dinner in peace. When the exposition was over, Mr. Black took up a collection on behalf of the heathen and all others who had not the happiness to live in Christian homes, after which we sang a final hymn and dashed out into the open air with great shouts of delight.

The monthly preaching was held at full moon, but it was in no sense a feast of Luna, for the choice of the date was determined solely by the state of the moor and woodland of the paths. It would have been difficult to imagine anything less like a pagan ecstasy or the divine irresponsibility associated with the moon. Thirty or forty people crowded on the long forms in the little room, a paraffin lamp stood on a table in a corner and the minister preached from the mantelpiece as pulpit. Most of the congregation were women; the rest were children, with a very few men; and the result was intolerably depressing. You came from the still frosty woods, where there was no sound but the burn telling its chaste numbers to the stars, into the fooshty air of the room where the smells of peppermints, paraffin and Sunday clothes blended into something like remote but powerful cheese. The pained women gossiped, desperately making the most of the few minutes before the service began. Their whispering voices hissed like snakes, their glasses shone malignantly and their thin heads made swift reptilian darts as if to strike their victims. A sudden and rather guilty silence fell as soon as the minister entered. Everyone became interested in their hymn-books and the more malicious women tried to look like wise virgins ready for the coming of the master. The service opened with a hymn through which Miss Sarah Todd led us with her amazing soprano which climbed higher and higher until it broke in a fearsome screech against the battlements of heaven. The rest of us had to follow her as best we might and forced our voices up so high that we felt as if we were trying to sing out of the tops of our heads. By the end of

three verses the praise had dwindled to a solo from Miss
Sarah accompanied by a hollow rumble from a few of the
men. But even Miss Sarah had her limitations, and her top
note in the next verse would be a shrill sound like tortured
glass, after which she would adroitly come down an octave
into the range of ordinary throats. As a display of virtuosity
her singing was worth the many discomforts of the preaching. I hated the service. It was dreary. There were too many
words and too little fresh air. I could not see any meaning
in it and I doubt if the congregation did either, but it was
something to attend on the long winter nights before the
W.R.I. brought the arts and crafts and community drama
to the countryside.

However the Christmas treat paid for all the discomforts of the Sunday school and the preaching. It at least I
enjoyed from beginning to end, for it was organized by my
grandmother, who knew the things that children liked. We
began with eating and drinking and passed on to games such
as ducking for apples where we could splash water about and
make an almighty noise. When we had taken the edge off our
spirits we had to entertain the older people with songs and
recitations, after which we divided the remains of the food
and staggered off for home. So it was on this last Christmas
treat nineteen years ago.

I have one memory, which I will never lose, of myself,
a fat, round child gorged with sweetmeats, standing in the
middle of the room in a Gordon tartan kilt and reciting a
poem that was rank with the fury of war. I have forgotten
all of it except two lines which, enunciated with a fine sense
of the dramatic, brought the house down:

> O Kaiser Bill, ye'll get yer sairin'
> In Hell they'll roast ye like a herrin'.

It was by far my most successful public appearance. I
had translated the patriotic feelings of the Sunday school

Christmas treat into words and I was rewarded with applause such as had never been heard in that room before. Even the minister whom I disliked, and who knew I disliked him, slapped me on the shoulder and gave me an apple, while Mr. Willie Todd remarked that I might be somebody yet. We then sang a song of peace and goodwill towards men and went home. We never met again. Mr. Black died a few days later and it was decided to suspend the school till after the war.

Everybody had forgotten about it by 1918 and it has never been revived. So another part of the old world passed away.

I have often wondered about the Old Man's attitude to the war, for it was self-contradictory in some ways and yet I suppose it was quite simple. I think he regarded the war as having nothing to do with him. He had not quarrelled with the Germans; they had never done him any harm nor had they even threatened him; therefore why should he want to fight them? At the same time he took a great interest in the progress of the fighting. He read every paper he could lay hands on, collected maps of all the battlefronts and regarded the loss of thousands with the impersonality of a High Command. It was a game in the grand manner played out in fields that were quite remote from his own.

He got some entertainment nearer home from the antics of the local defence force. This was a body of men who were too old or too decrepit to join the Army but who were formed into a last line of defence in case of invasion. They were a hundred strong and met every Saturday afternoon on the moor where they were drilled by a superannuated sergeant from the town. They were a noble sight as they formed up on the brow of the moor overlooking the sea to bid defiance to any German battleship that might lift above the horizon. Some were long, thin and hairy; some were short, fat and bald; some were bandy-legged, some intoed, and many were twisted with rheumatics. No possible arrangement could reduce them to anything approaching uniformity and their oddness was increased by the fact that they wore their ordinary clothes and carried a strange assortment of rifles. The sergeant was old, but time had not abated the fire of his temper or the fury of his language. If words could have made soldiers he would have sworn the defence force into a dandy regiment; but neither words nor passion could make those Scottish farmers forget their native thrawness. Not more than half of them obeyed an order at the same time and there were always a few incapable of distinguishing between right and left. Thus even the simplest manoeuvre such as forming fours landed the company in confusion and the sergeant in flaming incoherence. But he had the indomitable spirit of an Old Contemptible. Though they nearly broke his heart he put them through it. He drilled them till their shirts stuck to their backs, then he drove them on to field exercises such as advancing under cover across broken country. That was fun, for, at the order given, they had to flop on their bellies and creep from one whin bush to another. Their rheumatics gave them hell in the unaccustomed positions, whins lurked in the grass to torture them, and the sergeant lacerated them all the time with the rough edge of his tongue. At such times the Old Man would drive past in the gig and draw in to enjoy the sport. It must have given him deep satisfaction to see the pompous worthies of the neighbourhood falling on their pneumatic bellies and dodging about among the whins like rabbits with their broad bottoms to the sky, and a rare pleasure to hear them sworn at as they used to swear at their servants. He would sit in the gig looking at them with an expression of cynical amusement on his face, then he would drive up to the sergeant to pass the time of day.

'Brike yer bleedin' 'eart, they would,' the sergeant used to say, wiping the sweat from his brow.

'O, they're comin' on nae sae bad,' the Old Man replied, 'I've seen waur turns in the circus mony's the time.'

Then he would drive away home, more than ever convinced of the essential madness of the human race. He was a farmer and a man of peace.

But, if he had seen the Germans coming over the brow of the moor—that would have been another matter. He would have died sooner than let them take away Dungair. When the invasion scare reached its worst, we received orders for the evacuation in the case of a landing. The cattle were to driven to one mobilization point. The grain was to be sent to another. The families were to pack their valuables into the carts and drive to a third. We were to burn everything that we could not take with us. That appealed to the Old Man as a splendid game and he did his best to persuade Sandy to hold a rehearsal because, he said, there were a lot of things about the house that would be better to be burned, but Sandy was always too busy about other things so the Old Man had to content himself with perfecting his plans in his own head. He of course did not intend to leave Dungair. The others might go, but he would remain and defend it to the last. He and I made the preparations for the siege, assisted by the indomitable Captain Blades. We strengthened the arbour with sandbags, until it looked like a little fortress, then we provisioned it with water and rum and armed it with two double-barrelled guns and a few cases of my grandmother's ginger-beer as hand grenades. By the time we had finished it Captain Blades was so fired with the adventure that he offered himself for the garrison, an offer which the Old Man accepted. So the two old gentlemen prepared to defend their native land. I am not sure just how far the Old Man was in earnest. If the Germans had landed I think he would have carried out his plan; he was too old to go traipsing across the country and I am sure he could not think of life apart from his farm; but I also think that he never expected an invasion and that the fortifying of the arbour was a means of casting ridicule on the war hysteria of the countryside. Anyway it was good fun and we all enjoyed it.

The war went on and on as if it would never end. The flower of the young men who had sported at the games were dead. Boys were growing up to be soldiers. The values of the world were changing. There was a change working at Dungair too. Sandy's moods grew stranger every day. Sometimes he wanted to enlist but of course he was indispensable to the farm where he was doing two men's work. At other times he would be silent for days on end. It was obvious that he was suffering from a profound unease. Then the Old Man began to talk to him about marriage. I suppose he thought a wife would ensure the succession in Dungair. But Sandy just grunted and went about his work as morosely as before. Then he spoke his mind at last. He wanted to go abroad when the war was over; he was tired of Scotland, tired of Dungair. He wanted to see things and make big money. At the first chance he would go where his brothers and sisters had gone. And the sooner the better he would be pleased. That was all there was to be said about it. Perhaps the Old Man had suspected what was in the air. He told Sandy he was a damned fool and never mentioned the matter again, but he kind of lost interest in things and got ready to die.

We led the harvest of 1918 in fair order, but the weather was broken, our workers were indifferent and it was a sair struggle to get the crop home before the winter set in. The Old Man must often have thought of the golden harvest of 1914 and maybe he knew he would never see such another again. There was no time that year to bother about fine ricks, but the Old Man did his best to tidy the corn-yard with the rake and, though the glory had departed, he might still have taken some pleasure in the yard when all the stuff was home. But he was now feeling very old. A wound in his thigh, where a harrow had fallen on him forty years before, turned bad and gave him a great deal of pain. He could not sleep at night and life became a burden to him. He knew he was going to die soon. One afternoon at the end of October he yoked the pony into the gig and set out on a last visit to his friends. He looked in to take a farewell dram with his old acquaintances; then he made a round of the inns where he had so often had his pleasure. It was midnight, very cold and wet, when he came home to Dungair at last, sitting four-square as ever in the gig. He went to bed unaided and enjoyed his first sound sleep for weeks, but when he woke he had a fever which grew steadily worse towards night. His breathing became difficult and the pain in his thigh increased. His strength failed rapidly. He sent for his lawyer and made his will, then told my grandmother that he was going to die. He lasted for a few more days, becoming always weaker as the pain increased. My grandmother tended him unceasingly and though he sometimes swore at her he was not ungrateful. Yet he knew her labour was useless. Towards the evening of the last day he waved away the glass she was offering him.

'Spare yoursel,' he said. 'Throw the aul devil oot tae the midden and lat 'im rot in peace.'

Then he turned over to the wall and died an hour later. We buried him on the tenth of November in the kirkyard on the hill where the rest of his people were waiting for him. All the old men of the countryside attended the funeral, drank a dram to his memory and followed the coffin through the woods to its grave. A glim of sun shone as we planted him in the ground, but the wind blew up and rain was falling over the wintry landscape when we left him. Sandy and I drove home alone and both of us knew, though only half aware, that we had left more in the grave than the body of an old man. Dungair was silent and empty when we got home and we knew and my grandmother knew that it was no longer home.

The Armistice was signed next day and the intolerable folly of the war came to an end. A month later Sandy received an offer of a job in California. He accepted it; we gave notice that we would leave the farm in May; and my grandmother began to look for a house near the town where she could end her days in a garden by the river. We found just such a house, a red-tiled cottage sheltered by trees, with a long garden sloping down to the riverside. As the spring came in we sowed the last crop at Dungair, a crop we would never reap; my grandmother took over the cottage and began to prepare it for her ingoing; and we made arrangements for the sale. The farm had been let to a rag-and-bone merchant who had made a small fortune during the war and thought to become a gentleman farmer. He began to come about the place as if it were his already and, though we resented his manners, we could not deny him his rights, but he made us all the more anxious to be gone and it was with great relief that we saw the day of the sale draw near. And that day came, the unthinkable day that put a final end to our long association with Dungair.

We were sold up in the middle of May.

The auction began in the court at eleven when the auctioneer, having made a stock speech about the melancholy nature of the occasion, asked for bids for the first lot of odds and ends from the tool-shed. Those were the bargains that the Old Man had picked up at many a like displenishing and now they were to move on again. I doubt if we made a profit on them, for I am sure they did not fetch more than ten shillings, but they brought joy to the hearts of many old men who bought them in small lots at sixpence a time. It was not that the purchasers thought the things would ever be any use to them but they wished to have some memorial of the Old Man that they could show to their friends and say, 'Aye, that was aince Dungair's. Puir man, he's deid thae three year. There winna by mony o's left ere lang.' A pleasure cheaply bought at sixpence. We found better trade for the implements. Prices were high in 1919 and tools such as ploughs fetched more than they had cost when they were new before the war. The binder, the mower, the reaper, harrows, grubbers, ploughs, yokes, swingle-trees, socks, coulters—one by one they were knocked down and we took it as a kindly act when they were bought by friends. It was almost unbelievable the stuff that came out of corners of the steading—things so old that we had forgotten what they were used for. But no matter how decrepit or obsolete, every lot found a buyer and by mid-afternoon only the cattle remained to us. These went next. At five we owned nothing there but the clothes we stood in. At six my grandmother and I drove in a friend's car to the house at the riverside and that, as far as we were concerned, was the end of a long story which had taken over a hundred years in the telling.

For some others, however, it was not quite the end, though what remains has never been told and may never be, for none of those who played a part remember very much about it. But from exterior evidence this is a rough outline of what happened. The roup marked the end of a long tradition of hospitality and riotous living, therefore Sandy and his friends made up their minds that it should go out in style. We had laid in a big stock of drink to treat our friends that day. A good deal of it remained after the sale was over and that was augmented from the refreshment tent which an innkeeper had set up in the cart-shed. So, when the sale was over and the common people had gone, Sandy and his friends shut themselves into the empty drawing-room with a cold roast, a few loaves and lashings of drink. They ate, they drank, they sang and told stories while the evening declined into the dewy dusk of May. The uproar in the room must have been terrific as they revived the ancient glories of Dungair and celebrated the Old Man in yet wilder and more drunken toasts. Then someone discovered a bowl of eggs that had been overlooked in the dairy. A surfeit of remembrance awoke childish impulses in the old gentlemen. The sport turned into earnest and when they had plastered themselves with eggs they began to throw bottles. There is a divinity that protects drunk men, else some of them would have been killed; as it was, a few of them carried black eyes and ugly bruises for a long time in witness to the splore. They staggered home in the grey dusk of the morning, leaving the drawing-room a shambles of eggs and broken bottles. Every pane in the window was broken and the chandelier was smashed to atoms. No one who was there would say what had happened, but it was plain to see that we had ended as we lived, in riot.

There is no more left to tell, for the story of Dungair ended with the death of the Old Man. My grandmother was to enjoy the peace of her garden and the love of her friends for ten more years, and I think the happiness of those later days repaid her sorrows many times. But a story of such quiet love does not belong to Dungair where life was always too near the elemental to find the poise of beauty. I foreswore that life, yet I returned to it, for this one thing must be said of my fathers—that they were alive. They found something which they loved; they served it all their lives; and now that they are dead it remains as their memorial. If I could be as fortunate I would not have lived in vain.

Brig o' Don Boy

Unfinished sequel to Farmer's Boy

Brig o' Don Boy

The cottage was very small after the big farmhouse in which I had lived my first eleven years. There we had the drawing room with its ornate furniture and wallpaper starred with gold, the dining room where men sat round the table all night playing nap and drinking whisky, the parlour where my grandfather and I slept in the brass bedstead, the kitchen with its fireplace that took peats by the barrowful, the scullery with its fifty-gallon boiler for washing the milk cans, the laundry in which no one laundered, and the long narrow milk house where the basins of milk sat for the cream to rise. Upstairs there were two large rooms in front with bow windows and brass bedsteads, and two small rooms behind with iron beds and skylights. When I heard the minister on Sundays affirm that in Our Father's house there are many mansions, I knew they would be as spacious and familial as our old house, though it was unlikely there would be any nap played and spirits drunk there.

Our state had narrowed in the cottage, which had only the but, the ben and a little bedroom between. The kitchen had a range with an oven at the side in which a baking heat could be raised if the flues were cleaned. For more delicate cookery we had a paraffin stove, Rippingsville by name. It smelled a bit, which may have been the reason why we used onions with everything. About half the room was taken up with my grandmother's brass bedstead, a thing of many shining knobs that jangled like little bells if one pranced about. I could imagine myself as Santa Claus with reindeer, till I outgrew such fancies and started to follow girls, at a safe distance, in my mind.

The ben held the afterglow of our former state. It had the sideboard, the handsomest thing I had ever seen. It was not of mahogany, but light and inlaid. Its front was square, a narrow frame enclosing a fine sheet of glass, incised of course with delicate ferns and the like. Then at each side there was a little rounded cupboard with a rounded door of delicate wood – walnut, I think; it must have been walnut. Those cupboards held our small treasures of china – teapots in the shape of frosted ducks spangled with gold or silver or both, and the pig with the slit in his back which was my exchequer. On top there was a marble slab, like a great lovely piece of cheese. Two vases sat on it; or you might have called them large tureens. They were, I think, of china and were encrusted with leaves and flowers, touched with silver and gold. They were very rich, but I never liked them. They were wedding presents from the late '70s, when the rococo had declined into misplaced ingenuity. Somehow even in my earliest teens my taste was for an orderly style, though that may not have been shown in my conduct. The sideboard was topped off by a mirror in a gilt frame whose elegant shape was flawed with vine leaves and curlicues. But it was a fine enlarging glass that reflected all of the little room and doubled it. Next to the sideboard was the piano, a decent thing with only a few scrolls of inlay. Its keyboard was all invitation to me, and I could improvise on it with one finger or even two for long times, discovering wonderful relations of sound and interval until I was put to lessons from a hardworking teacher who soon made the instrument a torture for me. That was a bad time until I developed pains in my right hand which stopped the lessons. It must have been then that I discovered Psychosomatics. The third chief piece of furniture was a new thing got for our narrower living. It was a sofa which unfolded to become a bed. It may not have been very comfortable either way, and did not go very well with the sideboard, but it was handy. There were also an easy chair, a plain chest of drawers for putting photographs on, and a set of deal bookshelves to hold my few books that gradually became many.

The little room held a bed, a chair, a chest of drawers and myself, with just enough space to close the door. It was lit by a glass mantle on a gooseneck above the bed, the mantle being enclosed by a shade patterned, as everything seemed to be, with leaves and flowers. When I put out the light I was often joined by a mouse who did well, as I was always a crumby boy. As I never let on that I saw the mouse, thinking that he deserved his own quiet life as I deserved mine, he got confidence and came out in the gaslight. I liked him very much, for he used to sing in a tiny voice, notes in some quadruped scale. It was a very long time before anyone would believe me about that singing, but I did at last find a learned man who said yes, the mouse had some form of bronchitis. That I could believe, for I too had bronchitis in a chronic form and I too sang.

The small cottage was big enough for my grandmother and me, and I have never really wished for any more. However we could not have it all to ourselves all the time. My grandmother had a fortune of £900 and no hope of any more. This, she explained to me gently, as she explained everything to me who was her closest friend, must last us until I had got my education and was able to make my living. By that time she thought I would have only myself to care for because of her increasing infirmities, but hoped she would live just that long. She died more than a year too soon, but her good management took me on to what she had wished for. When I had graduated from the University and went to my first job in Glasgow at £2 a week, my trustees handed me the balance of the estate, which was £30: after ten years.

She could see only one way to eke out our capital and that was to take in a lodger, while she was strong enough to do so. A permanent one might have been a monotony: and besides, there was always the chance of bigger money in the holiday season. A student, she thought, would be ideal, some good industrious laddie who might help me with the Latin and French. One such did come our way, a very quiet and industrious young man. He was from fisher people, with a mother who might have been my grandmother's twin in all the simple verities and did become her friend. He did help me a little, and would have done more, but I was too shy to ask and I guess now he may have been to shy to offer. He was the quietest and pleasantest lodger we ever had, and we were grateful for him.

The lodger got the ben and the little bedroom and so, during his term, I was displaced to the attic. I thought that altogether romantic, which was fortunate, for it was dark and cold and bare. It was approached by a wooden ladder, up through a hole in the lobby roof, and somehow I always felt like Noah going up into the Ark. The roof rested on the top of the cottage walls, low pitched and perfectly triangular in section. The couples were floored over with deal boards and the rafters rose up from that floor at very acute angles, perhaps 45 degrees. The rafters were solid and rough, a hundred years old and cut with a pit saw. They were unlined, and carried the red brick tiles which may have been made at the Seaton brickworks where my Great Grandfather had worked as a boy in the beginning of the last century. The pattern of the tiles was more interesting than any boards could have been. The light came in through two of them, which were of thick glass moulded in tile shape. Other light came through where the cement had fallen out, along with the smell and the feel of the sea when the wind was in the east. The only window was a tiny one in the gable end: less a window than a porthole, from which I could look out on the river mouth and the traffic of ships in the bay.

The attic was too large and draughty for a bedroom, so we made as it might be a cabinet in the west end, under the two glass tiles. The dividing wall was a frame of laths, the kind used for making hen runs and the like, and this was covered with sacking, the very tough course linen stuff used for meal sacks. We covered the rafters too with old corn and fertiliser sacks, well tacked on but not very precisely, so that the arras bellied here and there. It was not like the flowered wallpapers to which I had been accustomed, with their great green lilies and cabbage roses and gaudy birds, but it had its own interest, of fading letters that read *Best Scotch Oats* and *John Milne's Superphosphate*. The furniture was a bed, a chair, a small table and an orange box, which held a few books and my candle. It may not seem much, but it was my own, and I decorated it according to my own interests.

The cottage was within sight and sound of Pittodrie, where Aberdeen football team played, a band of heroes such as had not been seen since the Knights of the Round Table fell one by one in Lyonesse about their lord. From the cottage I could see the ground a mile away; and hear of glory or anguish from the crowd as the tides of fortune rose, but more often ebbed. I worshipped those heroes of Pittodrie, and because of them all others such. The *People's Journal* printed photographs of all the Scottish teams. These I cut out and posted on the old sacks so that I had a gallery of young men who had the best of all fame: they played in the papers. I knew them all, could reel off the names as well as I could the twelve-times table. But the passage of time is a sorry thing, so no names stay with me now but Donald Colman and Napper Thomson. All the rest are gone with the wind off the North Sea.

I had another set of pictures cut from the newspapers, different but maybe not incongruous. You see I grew up in the days when we had farm horses, and grooms used to travel stallions for the service of the mares, giving, it was said, a bit of service themselves to handy girls in the by-going. Some of those Clydesdale stallions were as famous all over the Scottish countryside as the leading sires of racehorses in a different but not any more sophisticated society. They were the supreme attractions at the Highland and other shows, and their photographs appeared in all our papers. So my gallery included Baron of Buchlyvie and Dunure Footprint and . . . as with the footballers, the names have gone from me, but I can still see those forms of perfection and potency. And not only horses, but bulls of Collynie breeding and the quiet Aberdeen Angus with the broad hummel faces. All those went well enough with the football players. And of course were worth a great deal more in the market: an order which is now clean changed when champion footballers go for £100,000.

I could imagine that all the champions, beasts and men, were my own inside the space of my little Ark. I practised a great deal of make-believe there because my circumstances and my shy nature and my girth did not allow me the full life I would have liked, and of which I was sure I was capable. This make-believe went with me anywhere. In solitary walks by the river I was not alone but always playing a part in company – a corn merchant buying oats from an unwilling farmer, which I would sell at a grand profit. Or any other part that expressed myself. There was nothing fanciful in the deployment of those situations. I used the dialogue of everyday speech and argument back and fore as I was accustomed to it, whether I was the grain merchant after profit or Sir Lancelot after Queen Guinevere. I spoke the words in my head – I still do when I am writing – and often and often spoke them aloud. When I was overheard I was thought ridiculous, especially by those who found words hard to come by. They said I was a bit mad, and would have thrown old potatoes or horse dung at me. Wiser people just laughed and said it was maybe a sign that I would be a professor if I didn't land in the asylum first. They were not unkind, for there was an old respect for madness as for learning, both being afflictions, from God or the Devil, and either way, not to be mocked.

By the time I was thirteen or so, I had learned to keep those dialogues private and under my breath. The attic was the ideal place for them. My bedroom was divided off at the west end. The rest was bare to the tiles and dark, having no light but what came through the foot-square window in the gable. There was a variety of old things in the attic, trunks and stools and one of those headless bodies on which dressmakers used to fit their work in progress. With the trunks and stool I made a study for myself close up to the window, having the headless woman, dressed in a gown and dolman, as a silent companion. When the weather was not too cold I did my lessons there, but more often let my imagination work. The headless woman was not beguiling, though I already had an idea what girls were for and that it was pleasure. I had no doubt what one could be doing with them, though how to get doing it was far from clear to me.

The view from the small window was better. As it commanded the bay, there were always ships coming from or going to distant places. To one surrounded by the difficulties of ordinary life, there seemed nothing more desirable than to sail on one of those ships east and down and out of sight below the horizon, and after many days to make landfall by immemorial woods and be welcomed by some burgh of unimaginable freedoms. Or if that was excessive, to go fishing on the banks and catch netfuls of silver darlings, with the chance of a mermaid or a Rhine maiden at the new moon.

For pirates I had never much use, nor fairies even, which may have been the reason why I could see *Peter Pan*, when I saw it much later, without clapping a single hand.

About this time the place by the window was often wonderfully transformed. I became greatly interested in cricket. I had never played except when I was stood up before a wicket drawn on a wall drawn with chalk and tried to hit a soft ball with a roughly shaped stave. It was very seldom I hit the ball, for no one had told me I should watch it onto the bat, and that I sometimes survived an over or two was due to the bowlers being no more adept at hitting the wicket. At my school there were none of the playing fields that, as I understood, made Eton and Rugby what they were. I had therefore no chance to better myself, except by reading and in imagination. I read the reports of the English cricket matches in the papers and became as enthusiastic as about football. I was uplifted by the centuries made by Rhodes and Hobbs and Woolley, tried to see their strokes in my mind's eye and did see them. Not very well, perhaps, for when I did see Woolley playing his strokes there was a touch of difference. However what I made for myself was good enough for me; and of course I had to take part in the game. I could make up a report of a match, stroke by stroke, in which Hobbs and I made 349 against the vicious Australians, until I let myself be put out first, in proper deference to my companion and master. But that was too much make-believe; had not enough of the cruel blows of chance. Then, one good day, I fell on a device that put rules and bounds to what had been only indulgence. I used two packs of playing cards, and dealt out the cards in two piles, one for each batsman at the wicket. Each card had its face value in runs, up to the King which was worth 6 with 7 overthrows; a nice 13. A joker (there were four) meant the batsman was bowled, two of spades that he was caught and a two of clubs that he was stumped: the mischances thus being 8 in 104. That could be too hard odds, especially when Gubby Allen was batting, and refinements were introduced whereby on occasions an untimely joker or deuce did not count. I forget now how it worked, but it worked well for Allan, and I suppose there was sound reason for those let-offs. As I found in real cricket much later, the umpire does not always see everything and may even choose to be looking the other way. This was a fine private diversion for a long time. I filled yards of brown paper with scores of matches, and league tables and batting averages in which Allan was always well up there with Hobbs and Hearne and Woolley. I lost interest in this particular exercise after a season or two, but something of the kind stayed with me. For many years I put myself to sleep with interminable stories in which I was the hero: interminable because I was always asleep before I got to the end.

I was condemned to an academic education because I was not fit for anything else. I was considered unpractical and useless with my hands. Not wholly useless. I could be trusted to delve a yard and muck out a henhouse, but that was the limit. It was a long time before anyone encouraged me to think otherwise. Gradually I discovered I was not so very ham-handed and was not a complete fool with machinery. There were advantages in discovering such aptitudes so late. I had learned not to let on about them, which was wise. If you just stand around and look helpless when a machine breaks down, you can be sure that some practical person will come to your aid, probably making bad, worse. If you let it be known you are a handyman then you will be getting all sorts of fiddly jobs to do about the house. Therefore I have done my best to guard the illusion that I am just an intellectual fit to handle only a pen.

Those who had such a poor opinion of me may have had good reason for it. I remember for instance the dreadful things that happened in the woodwork class when I was thirteen or so. I was nervous of all the sharp tools we had to use: nervous, that is, in case I might damage them. My teacher had no time to give me any assurance, and the result was that I did damage them and myself. I cannot bear to think much about those hours of torture, made bearable only by the teacher kindly looking the other way. My best piece was the complete sum of misplaced endeavour. It was a pot stand, a little platform thing of cross-pieces fixed on two base blocks. The cross-pieces were ragged and unparallel. They were also stained with blood: my blood, and perhaps watered with a few tears. I think I could make a nice pot stand now, but I will not try it, for it might commit me to endless carpentry. I believe in the division of labour and I am against all that do-it-yourself – for myself at least.

Then there was the sad case of the bicycle which proved to me completely that in the handling of machinery I was a complete fool. The lesson was all the harder, because I thought a bike was one thing I could handle like a master. From the age of nine or so I had a green one that was a great joy to me. It was delightfully simple, having nothing more than the necessary wheels, pedals, handlebars and chain. It had no mudguards by the time I had learned to ride it after many upsets and collisions. It had the lower half of a bell, the top having been flattened by a road roller. I managed it very well, I thought, seldom running into the cows and sheep that were so common and yet so unexpected on the roads then. I even dared cycle with one hand in my jacket pocket. No hands would have been beyond my small courage.

When I went to school in the city, I had outgrown the green bike and wished for something better, to keep up with the West-End boys and their wonderful machines that had gas lamps and everything. My grandmother always gave me everything she could, but a new bike was beyond our income and I knew it. However I had capital – and I had it because I realised some money was not for spending, a piece of rudimentary economics I have somehow forgotten to use. I had two sorts of income, the pennies and the silver. Pennies were easy come and easy go, to be spent on grulsicks, a fine juicy word which describes caramels and liquorice and conversation lozenges. Silver money was not expendable, but had to be taken to school on Thursdays and put in the Bank. Somehow I did rather well in the silver market. When cattle dealers and the like came around, I usually managed to be at the clinching of the deal and a luckpenny came my way. Then there was an old gentleman who wore spats while riding his bicycle, and he used to give me half a crown, big money that was connected in my mind with the spats. Also elderly ladies used to exclaim upon my kilt and prettiness and used to give me sixpence in the King Street tram; but there are no trams nowadays, nor kilt nor prettiness. One way or another I had laid up a capital of £12, which was just figures in a book, with no practical meaning. But urged on by desire, I had a sudden revelation. The figures meant cash, and cash could mean a bicycle.

This was big business, and needed much discussion and negotiation. We took all the advice we could get, and there were plenty who were willing to give it, especially those who said it would be a waste of money and besides, I'd likely ruin the thing in a fortnight. The world was very full of discouragers in those days. I think they overdid it and fixed our resolution. We bought a handsome bike, a Raleigh, I'm sure, with a gas lamp and a whole bell and a three-speed gear. Even now I can remember the feeling of pride that bike gave me, and especially the three-speed gear. The machine was so splendid I could hardly make myself go on it and might not have done so, only I had to show it off. No hero of old ever rode a heavenly charger with more sense of glory than I did my Raleigh. I felt better than any old hero, for none had had a charger with a three-speed gear. You should have seen me ride up the Gallowgate and down the Spital, pedalling grandly and changing gear with a master hand.

That as you may guess was my undoing. Nobody had told me it was unwise to change gear while pedalling. After a day I realised there was some mechanical trouble. I took the bike to good old Mr Tom who mended things. He gave the wheels a twirl, shook his head and said, "Laddie, ye've ripped the guts out o' the gears. She'll need to be a common bike from now on."

She was never any more than that to me. Just a common bike and a warning that, in the matter of machinery, I was, as everybody said, a fool.

Glossary

arnott, a type of edible plant root

birse tea, tea laced with whisky
bluffert, a squall
brakes, bracken
breenge, to rush forward recklessly

chuckie, a chick
cornkist, a grain chest
couples, inclined rafters supporting a roof

dolman, woman's cloak with loose sleeves

ferlies, marvels
fraising, making fine speeches

greybeard, a stoneware jar for spirits

halflin, an adolescent youth; a farm worker
hirstle yont, to move along
hummel, without horns (of cattle)

ingle, a fireplace

kittle cattle, unmanageable or difficult creatures

lowse, to finish work (for the day)
lum, chimney

midse, a dividing furrow between two ploughed ridges

nowt, cattle

partans, crabs
potestatur, prime (of life)

ripe, to hunt through, plunder
rositty, full of rosin

sowens, a dish made from oat husks and fine meal steeped in water
speir, to ask
splore, a spree
stirk, a young bullock
stot, a young ox
sug, a slovenly person
sweir, loath, reluctant

taigle, disrupt, delay, loiter

whin road, a road made of stone

161